Elements, mixtures and compounds

Substances can be (1) pure **elements**, (2) pure **compounds** or (3) **mixtures** of elements or compounds.

Atoms

You can find out more about atoms on pages 4 and 5.

All substances are made of **atoms**. An individual atom is too small to see, so everything around you contains very many atoms.

If an element is divided into smaller and smaller pieces, an atom is the smallest part that can exist.

You can find out more about the periodic table on page 9.

1 Elements

An **element** is a substance that is made of only one sort of atom. Oxygen is an element because it contains only oxygen atoms.

There are about 100 different elements. Atoms of each element are given a chemical symbol.

Every symbol starts with a capital letter, often followed by a lower case letter. For example, N represents a nitrogen atom, but Na represents a sodium atom.

The elements are shown in the **periodic table**.

2 Compounds

Compounds form when atoms of different elements are combined during a chemical reaction. Compounds can separate back into elements only through chemical reactions.

The different elements combined in a compound are in fixed proportions, shown in the chemical formula.

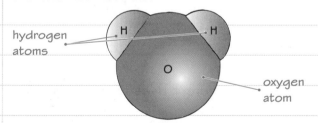

hydrogen atoms

oxygen atom

H_2O

The formula of every water molecule is H_2O because in each molecule two hydrogen atoms are chemically combined with one oxygen atom.

3 Mixtures

A mixture consists of two or more elements and/or compounds. These components are not chemically bonded. The chemical properties of each component in the mixture remain unchanged.

Worked example

Some iron powder and some sulfur powder are mixed, and tested with a magnet. The mixture is heated and cooled. The product is then tested with the magnet.

Explain why the magnet attracts some of the mixture before heating, but none of the product after heating. **(2 marks)**

The magnet attracts the iron, which is magnetic. When the mixture is heated iron sulfide is formed, which is non-magnetic.

In the mixture, iron retains its magnetic property, but the compound has different properties to the elements, iron and sulfur, from which it was formed.

Now try this

In the diagrams, each differently coloured circle represents the atom of a different element.

State which of the substances **A**, **B**, **C** and **D**:
(a) are pure
(b) contain no compounds
(c) could contain molecules of sulfur dioxide, SO_2.

(3 marks)

Filtration, crystallisation and chromatography

Mixtures are separated by **physical processes**. Separation of mixtures does not require **chemical reactions** and no new substances are made.

Practical skills: Filtration

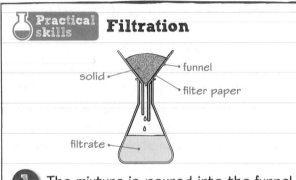

1. The mixture is poured into the funnel.

2. **Insoluble** solids remain in the filter paper.

3. **Solutions** and **liquids** go through into the flask.

Practical skills: Crystallisation

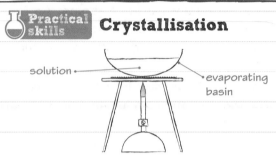

1. The solution is heated to concentrate the solution, and crystals may start to form.

2. It is then left in a warm place for the remaining water to **evaporate**.

3. **Crystals** form.

Practical skills: Paper chromatography

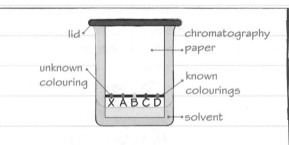

To set up paper chromatography:

- ✓ draw a pencil line near the bottom of the paper
- ✓ add spots of colourings to the line
- ✓ put the paper into a chromatography tank with a solvent at the bottom and below the line.

The **soluble** dyes are carried up the chromatography paper. Different dyes travel upwards at different rates. See page 74 for more about chromatography.

Worked example

The diagram shows the results of paper chromatography used to see if safe food colours are used in a sweet.

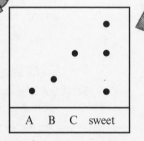

There are three spots above where the sweet's colouring was added.

(a) Explain which colours were detected in the sweet. **(2 marks)**

Colours A and C because they travelled the same distance as two spots from the sweet.

(b) Colours A, B and C are safe. Explain whether the sweet is safe to eat. **(2 marks)**

You cannot be sure because the sweet contained a colour that was not A, B or C, and that might be unsafe.

Now try this

Only sodium chloride dissolves in water.

You are given a white powder. It is a mixture of sodium chloride and calcium carbonate. Describe how you would separate the mixture to obtain pure samples of each substance. **(4 marks)**

Distillation

Distillation is used to separate a liquid from a mixture.

Simple distillation
Practical skills

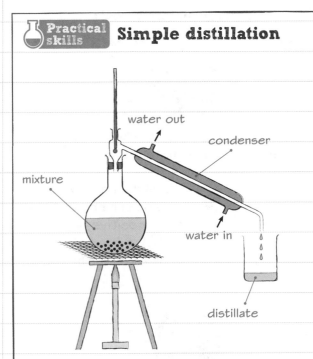

water out

condenser

mixture

water in

distillate

The condenser has two tubes, one inside the other. Cold water runs through the space between the two tubes, keeping the condenser cold. The cooling water does not mix with the substance being separated.

In a mixture of two liquids, the liquid with the lower boiling point is collected as the distillate.

In a solution of a solid in water, the water is collected as the distillate.

Fractional distillation
Practical skills

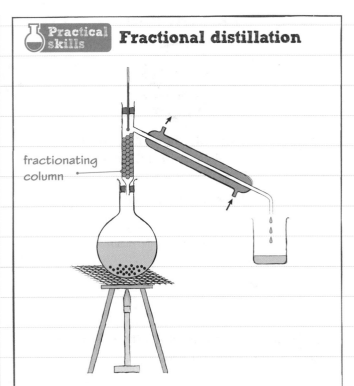

fractionating column

Fractional distillation is used to separate a mixture of more than two liquids using the laboratory apparatus shown.

The column is packed with glass beads. The different liquids leave the top of the column in order of increasing boiling point.

You can find out how crude oil is separated into fractions in industry on page 62.

Worked example

A mixture of ethanol and water is distilled. Explain a safety precaution that should be used. **(2 marks)**

Ethanol is flammable, so use an electrical heater rather than a Bunsen burner.

Practical skills Ethanol has the lower boiling point. This will boil off first.

Now try this

1 Explain the function of a condenser in distillation. **(2 marks)**

2 A mixture of ethanol (boiling point 78 °C) and water is distilled until no liquid remains. Explain what readings are seen on the thermometer.

(3 marks)

Historical models of the atom

As scientists discovered new evidence, the scientific model of the atom was updated or replaced.

1 An early model

Atoms were originally thought to be tiny spheres that could not be made smaller.

Electrons were discovered.

2 Plum pudding model

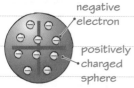

The atom is a positively charged sphere containing electrons.

The alpha particle-scattering experiment showed that most positive alpha particles fired at an atom went through the atom.

3 Nuclear model

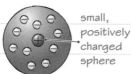

Atom's mass mainly a positive central nucleus.

4 Niels Bohr's model

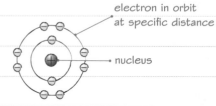

Bohr's calculations agreed with experiments showing the electron orbits.

Experiments showed that the positive nucleus was made of small, positively charged particles, which were given the name 'protons'.

5 Atomic model with protons

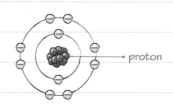

6 Atomic model with neutrons

James Chadwick's experiments showed, about 20 years after the nucleus had been accepted, that the nucleus also contained particles with no charge; these were given the name 'neutrons'.

Particles in an atom

Protons, neutrons and electrons are called **subatomic particles**. You can work out how many of each type of subatomic particle an atom has from its atomic number and mass number.

Atomic number

The number of protons in an atom of an element is called its **atomic number**.

The atoms of different elements have different numbers of protons – no two elements can have the same atomic number.

Number of electrons

Atoms have no overall charge. This is because the number of electrons in an atom is the same as the number of protons.

Mass number

The total number of protons and neutrons in an atom is called its **mass number**.

Worked example

A sodium atom has an atomic number of 11 and a mass number of 23. How many of each type of subatomic particle does it have?   **(3 marks)**

number of protons = 11

number of electrons = 11

number of neutrons = 12

The number of protons is given by the atomic number, which is 11 for sodium.

The number of electrons in an atom is the same as the number of protons, which is 11 here.

The number of neutrons equals the mass number minus the atomic number.

So the number of neutrons = 23 − 11 = 12.

Diagrams of atoms

Atoms of elements can be drawn like the one in this diagram. This means that it is easy to work out how many subatomic particles there are.

In this particular diagram, the crosses represent electrons. They are arranged around the central nucleus.

If the atom has three electrons, it must also have three protons. This means that the white circles represent protons here.

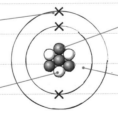

The black circles here represent neutrons.

The atomic number is 3 and the mass number is 7 (3 protons + 4 neutrons).

Now try this

The table gives information about two atoms.

Atom	Mass number	Atomic number
X	40	20
Y	40	19

(a) Calculate the number of protons, neutrons and electrons in atom Y.   **(3 marks)**

(b) Explain how you would know that X and Y are atoms of different elements.   **(2 marks)**

Atomic structure and isotopes

Atoms of the same element that have different numbers of neutrons are **isotopes**. The average value of the atomic masses of these isotopes is the **relative atomic mass**.

Atoms

Atoms can be represented like this:

mass number ——— $^{23}_{11}$Na ——— chemical symbol
atomic number ———

This is the full symbol for one atom of sodium, Na.

The alpha particle-scattering experiment is described on page 4, showing that the nucleus was a tiny part of the atom.

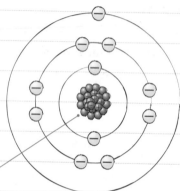

The nucleus has a radius of less than $\frac{1}{10000}$th of the radius of the atom, 0.00001 nm.

The table shows the relative charges and masses of the particles in an atom.

Name of particle	Relative charge	Relative mass
proton	+1	1
neutron	0	1
electron	−1	very small

Atoms are very small, with a radius of about 1×10^{-10} m (0.1 nm).

Almost all of the mass of the atom is in the nucleus.

Maths skills: Length conversion

$$1 \text{ nm} = 0.000\,000\,001 \text{ m}$$
$$= 1 \times 10^{-9} \text{ m}$$

Worked example

Two atoms of chlorine are shown $^{35}_{17}$Cl and $^{37}_{17}$Cl

(a) Explain why these atoms are described as **isotopes**. **(3 marks)**

The atoms are isotopes because they are atoms of the same element with the same number of protons and electrons, but different numbers of neutrons.

(b) Calculate the relative atomic mass of chlorine if it consists of 75% chlorine-35 and 25% chlorine-37. **(2 marks)**

relative atomic mass = $(\frac{75}{100} \times 35) + (\frac{25}{100} \times 37) = 35.5$

Isotopes of an element will have the same atomic number but different mass numbers. Isotopes of a particular element react in the same way because they have the same electronic structure.

Maths skills The relative atomic mass takes account of the abundance of each isotope. Multiply each mass by the percentage.

Now try this

1 Three atoms are represented in these diagrams.

In terms of subatomic particles, explain why they are isotopes of the same element. **(3 marks)**

2 Oxygen-16 and oxygen-18 are two isotopes. They can be shown as $^{16}_{8}$O and $^{18}_{8}$O.
(a) Describe **one** way in which the isotopes differ from each other. **(1 mark)**
(b) Describe **two** ways in which they are similar. **(2 marks)**

Electronic structure

You should be able to represent the electronic structure of the first 20 elements.

Energy levels

The electrons in an atom occupy different **energy levels** around the nucleus. Each electron in an atom is at a particular energy level. Electrons occupy the lowest available energy levels.

Shells

Energy levels are also called **shells**:
- The innermost shell is the lowest energy level.
- The outer shell is the highest occupied energy level.

Writing electronic structures

Different energy levels can contain different numbers of electrons. For the first 20 elements (hydrogen to calcium):

Energy level	Number of electrons
first	1 or 2
second	up to 8
third	up to 8
fourth	up to 8 (1 (K) or 2 (Ca))

For example, a sodium atom has 11 electrons:
- 2 fit into the first energy level
- 8 fit into the second energy level
- 1 fits into the third energy level.

This electronic structure is written as 2,8,1 (the commas separate each energy level).

Use your periodic table to work out electronic structures. Count from hydrogen to the required element, for example sodium.

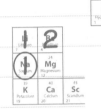 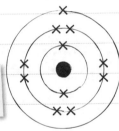

Electronic structures as diagrams

This is the electronic structure of sodium as a diagram.

Worked example

(a) Write down the electronic structure of oxygen.

2,6

(b) Complete the diagram to show the electronic structure of oxygen (atomic number 8).

(2 marks)

Make sure you can write and draw electronic structures correctly. The periodic table on the Data Sheet can help you.
- The number of energy levels (or circles) must be the same as the row the element is in.
- The total number of electrons must be the same as the element's atomic number.
- The number of electrons on the outermost shell must be the same as the element's group number.

Now try this

1 Write down the electronic structures for atoms of the following elements:
   (a) carbon (atomic number 6)   **(1 mark)**
   (b) sulfur (atomic number 16)   **(1 mark)**
   (c) calcium (atomic number 20).   **(1 mark)**

2 Complete the diagram to show the electronic structure of aluminium (atomic number 13).   **(2 marks)**

Development of the periodic table

Early scientists discovered many of the elements, and attempts were made to classify these elements. Mendeleev was one of these nineteenth-century scientists.

Information available in the 19th century

In the 19th century, scientists knew:

👍 the properties of the elements

👍 the **atomic weight** of the elements.

They did not know about:

👎 protons, neutrons and electrons

👎 atomic numbers (proton numbers).

> Atomic weight is similar to relative atomic mass, but not all of the atomic weights calculated at the time were correct.

An early periodic table

This early table:
- listed elements in order of atomic weight
- had seven columns.

This table was not successful overall:

👍 Counting along, every seventh element has similar properties.

1	2	3	4	5	6	7
H	Li	Be	B	C	N	O
F	Na	Mg	Al	Si	P	S
Cl	K	Ca	Cr	Ti	Mn	Fe

👎 The pattern failed after calcium.

👎 Some metals and non-metals were in the same column, such as oxygen and iron.

👎 There was no room for new elements.

Worked example

Dmitri Mendeleev published a table in 1869. The elements are shown in order of atomic weight. Part of his table is shown on the right.

Mendeleev left gaps in his table (shown as *). Explain why he did this. **(2 marks)**

Mendeleev realised that some elements had not been discovered, so left some spaces so that elements with similar properties lined up in groups.

Group

1	2	3	4	5	6	7
H						
Li	Be	B	C	N	O	F
Na	Mg	Al	Si	P	S	Cl
K Cu	Ca Zn	* *	Ti *	V As	Cr Se	Mn Br
Rb Ag	Sr Cd	Y In	Zr Sn	Nb Sb	Mo Te	* I

> The modern periodic table contains over 100 elements but far fewer were known in 1869. It is called the periodic table because elements with similar properties appear periodically (are found at regular intervals).

> Iodine has a lower atomic mass than tellurium so should come first. Mendeleev swapped them round to match their properties better.

> Mendeleev predicted properties for the undiscovered elements, for which he had left gaps. When these elements were discovered, their properties closely matched Mendeleev's predictions.

Now try this

1 (a) Which group is missing from Mendeleev's table? **(1 mark)**

(b) Give **one** similarity, and **one** difference, between Mendeleev's Group 1 and Group 1 in the modern periodic table. **(2 marks)**

2 Suggest **two** reasons why some scientists thought Mendeleev's table was **not** correct. **(2 marks)**

The modern periodic table

After sub-atomic particles were discovered, the arrangement of the elements in the periodic table was seen to be in order of increasing atomic (proton) number.

group numbers

1	2						3	4	5	6	7	0

		1 **H** 1

Key

relative atomic mass
atomic symbol
atomic (proton) number

| 7
Li
3 | 9
Be
4 |
| 23
Na
11 | 24
Mg
12 |

| 11
B
5 | 12
C
6 | 14
N
7 | 16
O
8 | 19
F
9 | 20
Ne
10 |
| 27
Al
13 | 28
Si
14 | 31
P
15 | 32
S
16 | 35.5
Cl
17 | 40
Ar
18 |

| 4
He
2 |

39 K 19	40 Ca 20	45 Sc 21	48 Ti 22	51 V 23	52 Cr 24	55 Mn 25	56 Fe 26	59 Co 27	59 Ni 28	63.5 Cu 29	65 Zn 30	70 Ga 31	73 Ge 32	75 As 33	79 Se 34	80 Br 35	84 Kr 36
85 Rb 37	88 Sr 38	89 Y 39	91 Zr 40	93 Nb 41	96 Mo 42	(98) Tc 43	101 Ru 44	103 Rh 45	106 Pd 46	108 Ag 47	112 Cd 48	115 In 49	119 Sn 50	122 Sb 51	128 Te 52	127 I 53	131 Xe 54
133 Cs 55	137 Ba 56	139 La 57	178 Hf 72	181 Ta 73	184 W 74	186 Re 75	190 Os 76	192 Ir 77	195 Pt 78	197 Au 79	201 Hg 80	204 Tl 81	207 Pb 82	209 Bi 83	(209) Po 84	(210) At 85	(222) Rn 86
(223) Fr 87	(226) Ra 88	(227) Ac 89	(261) Rf 104	(262) Db 105	(266) Sg 106	(264) Bh 107	(277) Hs 108	(268) Mt 109	(271) Ds 110	(272) Rg 111	(285) Cn 112	(286) Uut 113	(289) Fl 114	(289) Uup 115	(293) Lv 116	(294) Uus 117	(294) Uuo 118

The elements that Mendeleev swapped in his periodic table, based on atomic weights (see page 8), were in the correct order when arranged by atomic number.

Metals

- Atoms have a few electrons in their outer shell.
- Atoms lose electrons to form positive ions in ionic compounds.
- All metals except mercury are solid at room temperature.
- All metals conduct electricity.

Non-metals

- Atoms have a few electrons short of a full outer shell.
- Atoms gain electrons to form negative ions in ionic compounds or share electrons in covalent compounds.
- Some non-metals are made of simple molecules and are gases, liquids or solids with a low melting point at room temperature.
- Some non-metals have giant structures and are solids with a high melting point.
- Non-metals do not conduct electricity, except some forms of carbon.

Groups

Elements in the same group:
- have the same number of outer electrons
- have similar chemical properties.

Revise groups 0, 1 and 7 on pages 10–12.

Worked example

Tellurium and iodine have the symbols $^{128}_{52}$Te and $^{127}_{53}$I. Mendeleev put most of the elements in order of atomic weight but he put tellurium before iodine (even though its atomic weight is greater than iodine's atomic weight). Suggest a reason why he did this, and why it is not a problem in the modern periodic table. **(3 marks)**

Mendeleev realised that the order of atomic weight, iodine before tellurium, could not be correct because of the elements' properties. The correct order is in order of atomic number, tellurium before iodine.

Now try this

A new element was discovered in 2000. Livermorium has the symbol Lv. Its atomic number is 116 and its electronic structure is 2,8,18,32,32,18,6. Explain in which group of the periodic table Lv should be placed. **(2 marks)**

The **electronic structure** of Lv looks complicated but the last number shows the group into which it should go.

Group 0

The elements in Group 0 of the periodic table are called the **noble gases**.

The first three elements

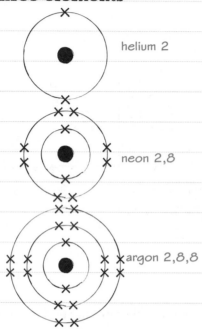

helium 2

neon 2,8

argon 2,8,8

The next three elements are krypton, xenon and argon.

Electronic configuration and reactivity

The elements in Group 0 include helium, neon and argon. The highest occupied energy levels (outer shells) of their atoms are full:

- Helium has two electrons in its highest occupied energy level.
- The others all have eight electrons in their highest occupied energy levels.

This stable arrangement of electrons means that the atoms do not easily form molecules and so are very unreactive (**inert**).

Boiling points

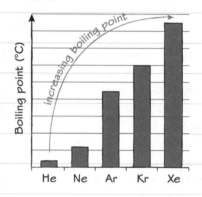

Worked example

In 1933 it was predicted that elements lower down in Group 0 might form a few compounds. Several hundred compounds of xenon, such as XeF_2 and XeO_4, have been formed in the laboratory. A few krypton compounds have been synthesised. The first argon compound was found in 2000. No helium and neon compounds are known.

(a) Explain why so few compounds are known of Group 0 elements. **(2 marks)**

Group 0 elements have atoms with full outer shells. This is a stable arrangement of electrons, so there is no need for these atoms to gain electrons, lose electrons or share electrons.

(b) What pattern is shown in this information? **(1 mark)**

The elements become more reactive going down the group.

(c) State whether you would expect the type of bonding in XeF_2 and XeO_4 to be covalent, ionic or metallic. Give a reason for your answer. **(2 marks)**

Covalent, because the atoms are all of non-metal elements.

Now try this

Krypton has an atomic number of 36. The most likely electronic arrangement of a krypton atom is:

- A 2,8,8,18
- B 9,9,9,9
- •C 2,8,8,8,8,2
- D 2,8,18,8 **(1 mark)**

You do not need to know the electronic configuration of krypton, but you should be able to select the correct answer by thinking about the outer shell of a krypton atom.

Group 1

The elements in Group 1 are called the **alkali metals**.

Properties of Group 1

The alkali metals have these properties. They:
• are metals
• have a low **density** (the first three are less dense than water so they float)
• form ions with a charge of +1
• react with water to form hydrogen and hydroxides
• react with oxygen to form solid, white, ionic oxides
• react with chlorine to form solid, white, ionic chlorides.

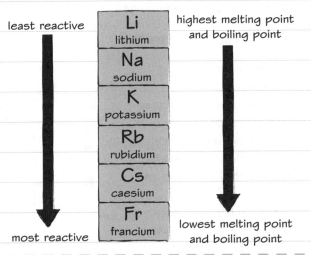

least reactive — Li lithium, Na sodium, K potassium, Rb rubidium, Cs caesium, Fr francium — most reactive

highest melting point and boiling point → lowest melting point and boiling point

Reactivity

The elements in Group 1 all have just one electron in their highest occupied energy level (outer shell) of their atoms.

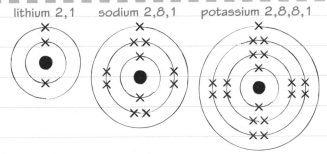

lithium 2,1   sodium 2,8,1   potassium 2,8,8,1

1 In reactions, the outer electron is lost, forming an ion with a 1+ charge.

2 All atoms of Group 1 have one outer electron, so they all react similarly.

3 These outer electrons are more easily lost going down the group.

4 So, the element's reactivity increases down the group.

Reactions with water

In general:

metal + water → metal hydroxide + hydrogen

For example, for sodium:

sodium + water → sodium hydroxide + hydrogen

$2Na(s) + 2H_2O(l) \rightarrow 2NaOH(aq) + H_2(g)$

Worked example

Sodium and chlorine react together quickly.
(a) Write a balanced equation for the reaction.   **(2 marks)**
$2Na + Cl_2 \rightarrow 2NaCl$

(b) How would you expect potassium to react with chlorine?   **(2 marks)**

Potassium would react very quickly to form potassium chloride.

Now try this

1 Room temperature is around 20 °C. Sodium melts at 98 °C. Suggest and explain a temperature for the melting point of potassium.   **(3 marks)**

2 (a) Balance this chemical equation, and include the appropriate state symbols:
$Rb + H_2O \rightarrow RbOH + H_2$ **(2 marks)**

(b) Give **two** observations that would be the same for the reaction in (a) and the reaction of lithium with water, and **one** observation that would be different.   **(3 marks)**

Group 7

The elements in Group 7 are called the **halogens**. The atoms of all the halogens have seven electrons in their outer shell, so the elements have similar reactions.

Properties of Group 7

The halogens:

- are non-metals consisting of molecules with pairs of atoms
- react with metals to form ionic compounds, forming **halide ions** with a charge of −1
- react with other non-metals to form simple molecular covalent compounds.

most reactive

least reactive

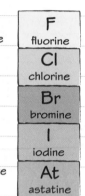

F	fluorine
Cl	chlorine
Br	bromine
I	iodine
At	astatine

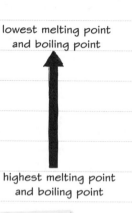

lowest melting point and boiling point

highest melting point and boiling point

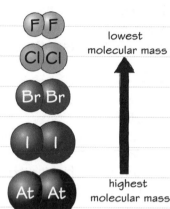

lowest molecular mass

highest molecular mass

The direction of these trends is opposite to the ones in Group 1.

Displacement reactions

A more reactive halogen can **displace** a less reactive halogen from an aqueous solution of its salt. The table shows the results of adding halogens to solutions of halide ions.

		Halide ion in solution			Number of reactions
		chloride, Cl⁻	bromide, Br⁻	iodide, I⁻	
Halogen	chlorine, Cl_2	✗	turns orange	turns brown	2
	bromine, Br_2	✗	✗	turns brown	1
	iodine, I_2	✗	✗	✗	0

bromine formed

iodine formed

For example:

Chlorine is more reactive than bromine so it displaces bromine from a solution of its salt:

chlorine + potassium bromide → potassium chloride + bromine

$$Cl_2(aq) + 2KBr(aq) \rightarrow 2KCl(aq) + Br_2(aq)$$

Worked example

When chlorine is bubbled through a colourless solution of potassium iodide, a brown colour forms in the solution. Explain why. **(3 marks)**

Chlorine is more reactive than iodine. The chlorine displaces the iodide ions, forming iodine. Iodine causes the brown colour.

The halogens get more reactive as you go **up** the group, so chlorine is more reactive than iodine.

Now try this

1 Iodine is a solid at room temperature. Explain which state astatine will be found in at room temperature. **(2 marks)**

2 (a) Describe what you **see** when bromine is added to potassium iodide solution. **(1 mark)**

   (b) Explain why this happens and write a word equation for the reaction. **(3 marks)**

Transition metals

The **transition metals**, including Cr, Mn, Fe, Co, Ni and Cu, have different properties to the alkali metals in Group 1.

Properties of transition metals

Compared with the metals in Group 1 (such as sodium), the transition metals:

- have higher **densities**
- have higher melting points and boiling points (except for mercury, which is a liquid at room temperature)
- are stronger and harder
- are much less reactive with water, oxygen and halogens.

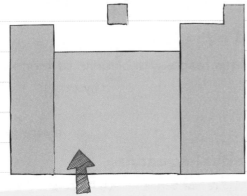

The transition metals are placed between Groups 2 and 3 in the periodic table.

More properties

Many transition metals also:

- form coloured compounds
- are useful as **catalysts**
- form more than one type of ion.

The alkali metals form colourless or white compounds.

Catalysts alter the rate of a reaction without being used up during the reaction.

Iron, for example, can form iron(II) ions, Fe^{2+} or iron(III) ions, Fe^{3+}.

Worked example

A red–brown oxide of iron has the formula Fe_2O_3.

(a) Explain how you can tell from this information that iron is a transition metal. **(2 marks)**

Transition metals form coloured compounds and iron oxide is described as red–brown.

(b) Give the formula of the iron ion in the oxide. **(1 mark)**
Fe^{3+}

Copper, for example, is another transition metal. It also forms coloured compounds such as copper(II) sulfate, which is blue in solution.

Oxide ions have the formula O^{2-}. There are three oxide ions in Fe_2O_3 with a total charge of $-2 \times 3 = -6$. The charge on the iron ion must be $+6 \div 2 = +3$.

Now try this

1 State **two** properties of copper that makes it suitable for making saucepans. **(2 marks)**

2 Aluminium and titanium resist corrosion, whereas steel does not.
   The table shows some of their other properties.
   (a) Name the densest metal. **(1 mark)**
   (b) Explain **one** advantage and **one** disadvantage of using aluminium instead of steel for making car body panels. **(4 marks)**

Metal	Density (g/cm³)	Cost (£/kg)	Strength (MPa)
aluminium	2.7	2.40	250
titanium	4.5	6.80	1000
low-carbon steel	7.8	0.85	780

   (c) Compare the use of aluminium with titanium for making car body panels. **(4 marks)**

Chemical equations

The **reactants** and **products** in a chemical reaction can be shown in a **word equation** or a **chemical equation** that uses symbols.

Word equation

Hydrogen reacts with chlorine to form hydrogen chloride:

hydrogen + chlorine → hydrogen chloride

reactants product

Chemical equation

There are 2 HCl molecules.

A hydrogen chloride molecule has 1 H atom and 1 Cl atom.

$$H_2 + Cl_2 \rightarrow 2HCl$$

A hydrogen molecule has 2 H atoms.

A chlorine molecule has 2 Cl atoms.

There are 2 H atoms on each side of the equation.

There are 2 Cl atoms on each side of the equation.

In a chemical reaction there are the same number of atoms of each element in the reactants and in the products.

State symbols

Sodium carbonate reacts with sulfuric acid, forming sodium sulfate, carbon dioxide and water:

$$Na_2CO_3(s) + H_2SO_4(aq) \rightarrow Na_2SO_4(aq) + CO_2(g) + H_2O(l)$$

solid — Sulfuric acid is a solution. — Sodium sulfate is soluble and forms a solution. — gas — liquid

> There is no need to add state symbols unless you are asked for them in the question.

Worked example

Write the balanced chemical equation for the complete combustion of propane, C_3H_8. **(3 marks)**

$$C_3H_8 + 5O_2 \rightarrow 3CO_2 + 4H_2O$$

> You can revise the products from the combustion of alkanes on page 63.

There are 3 carbon atoms in a molecule of propane, forming 3 carbon dioxide molecules.

There are 8 hydrogen atoms in a molecule of propane, so 4 water molecules are formed (each water molecule contains 2 hydrogen atoms).

There are 6 oxygen atoms in the carbon dioxide molecules and 4 in the water molecules, giving 10, so 5 oxygen molecules are needed (each oxygen molecule contains 2 oxygen atoms).

Now try this

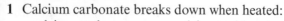

1 Calcium carbonate breaks down when heated:

    calcium carbonate → calcium oxide + carbon dioxide
       50 g   22 g

Calculate the mass of calcium oxide made from 25 g of calcium carbonate. **(2 marks)**

2 Explain, in terms of the number of atoms, why this is a balanced chemical equation. **(2 marks)**
   $H_2 + F_2 \rightarrow 2HF$

Extended response – Atomic structure

In each paper there will be one or more extended response questions worth six marks. In these questions you need to explain carefully all your knowledge and understanding, using any information given in the question. You must give your answer in clear English using scientific words appropriately.

You can revise the information for this question, which is about **atomic structure** and **isotopes**, on pages 5, 6 and 7.

Worked example

Magnesium exists as three stable isotopes, shown in the table.

Isotope	Abundance (%)
^{24}Mg	79.0
^{25}Mg	10.0
^{26}Mg	11.0

(a) Explain any similarities and differences that you would expect in the atoms of these isotopes and in the chemical reactions of **pure** samples of each of these isotopes. **(6 marks)**

(b) Calculate the relative atomic mass of magnesium to 1 decimal place. **(3 marks)**

(a) The atomic number of magnesium is 12, so atoms of ^{24}Mg have 12 protons, 12 electrons and 12 neutrons.

Atoms of ^{25}Mg have 12 protons, 12 electrons and 13 neutrons.

Atoms of ^{26}Mg have 12 protons, 12 electrons and 14 neutrons.

The atoms have the same number of electrons and protons, but different numbers of neutrons.

When magnesium atoms react, they will form a full outer shell. As they have two electrons in their outer shell, each atom will lose two electrons, forming a Mg^{2+} ion. This means that the chemical reactions of pure samples of each isotope will be the same.

(b) Relative atomic mass

$$= (24 \times \frac{79.0}{100}) + (25 \times \frac{10.0}{100}) + (26 \times \frac{11.0}{100})$$

$$= 24.32$$

$$= 24.3$$

> The question asks about the atoms and the reactions, so make sure that you cover both points.

> Use the information that you are given. The data given in the table can be used with the atomic number from the periodic table to calculate the number of electrons, protons and neutrons. This is better than making just a general statement about isotopes. A periodic table will be given to you with the exam paper.

> **⊞ Maths skills**
> 1 To calculate the relative atomic mass, multiply each relative mass (24, 25, 26) by its percentage abundance.
> 2 Remember that % (per cent) means /100.
> 3 The question asks for 1 decimal place, so round your answer. In this case the second decimal place number is 2, so the first decimal place number 3 is not changed.

Now try this

Describe how the electrons are arranged in a sodium atom, a chlorine atom and an argon atom. Use these electron arrangements to explain whether each element will react by its atoms gaining electrons, losing electrons, sharing electrons or not reacting at all. **(6 marks)**

> Give the electronic configurations of an atom of each of these elements. Remember that atoms normally form full outer shells when they react. In your answer give details about how each atom forms a full outer shell (if they need to at all). You will also find the next page useful.

Forming bonds

The noble gases are unreactive, but atoms of other elements form bonds. There are three types of strong chemical bond.

1 Ionic bonds

When metals react with non-metals:
- atoms of the metal lose their outer electrons
- these electrons are transferred to the non-metal atoms
- the atoms form oppositely charged ions with full outer shells

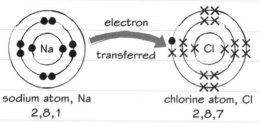

sodium atom, Na
2,8,1

chlorine atom, Cl
2,8,7

- the ionic bonds between the ions are strong (see also page 17).

2 Covalent bonds

When non-metal atoms are joined to other non-metal atoms:
- the atoms share electrons
- usually, the atoms now have a full outer shell
- covalent bonds are found in non-metal elements and non-metal compounds

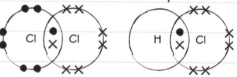

- the covalent bond between the atoms that share electrons is strong (see also page 19).

3 Metallic bonds

In metals:
- the atoms lose their outer shell electrons, forming ions
- the ions are in a regular, 3-D arrangement: a **lattice**
- the electrons from the outer shells move throughout the lattice
- metallic bonding occurs in metals and alloys

- the metallic bonds between the ions and the electrons are strong (see also page 24).

Naming compounds

H_2S

This substance contains hydrogen and sulfur only: it is named hydrogen sulf**ide**.

KNO_3

This substance contains potassium, nitrogen and oxygen: it is named potassium nitr**ate**.

You should be able to use the names and symbols of the first 20 elements, the compounds made from these elements, and of all the elements in Groups 1 and 7.

Worked example

Give evidence that shows the type of bond formed in the compound between the phosphorus and chlorine shown.

(2 marks)

Phosphorus and chlorine are both non-metals and they share electrons to form a molecule, so the bonding is covalent.

Now try this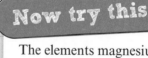

The elements magnesium and nitrogen react with each other.
(a) Write the word equation for the reaction.

(1 mark)

Remember that compounds between **two** elements are named ending in -ide.

(b) Explain how the bonds form in the compound.

(4 marks)

Ionic bonding

Ionic bonds are strong chemical bonds that form when a metal reacts with a non-metal.

Positive ions

Metal atoms and hydrogen atoms lose electrons to form positive ions.

For example, a sodium atom loses its outer electron to form a sodium ion, Na^+.

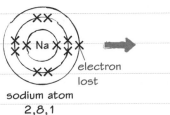

sodium atom
2,8,1

sodium ion
2,8

Negative ions

Non-metal atoms gain electrons to form negative ions.

For example, a chlorine atom gains one outer electron to form a chloride ion, Cl^-.

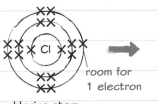 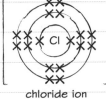

chlorine atom
2,8,7

chloride ion
2,8,8

Electron transfer

electrons transferred

magnesium atom, Mg
2,8,2

oxygen atom, O
2,6

magnesium ion, Mg^{2+}
2,8

oxide ion, O^{2-}
2,8

Ions and the periodic table

The atoms in Groups 1 and 2 lose electrons and those in Groups 6 and 7 gain electrons to form ions with a full outer shell (noble gas structure).

Metals		
Group 1	lose 1 electron	form 1+ ions
Group 2	lose 2 electrons	form 2+ ions
Non-metals		
Group 6	gain 2 electrons	form 2− ions
Group 7	gain 1 electron	form 1− ions

Noble gas structure

Ions have the stable **electronic structure** of a noble gas (an element from Group 0). Noble gas atoms have eight electrons in their outer shell, except for helium, which has only two. These atoms have no tendency to lose or gain electrons.

 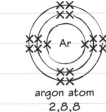

neon atom
2,8

argon atom
2,8,8

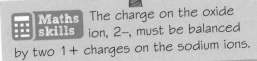
Now try this

(a) Describe how magnesium and chlorine form ions when they react to make magnesium chloride. **(4 marks)**

(b) Give the formula of magnesium chloride. **(1 mark)**

State the number of electrons involved when each atom forms an ion, and the charge on each ion.

Giant ionic lattices

In an ionic compound there is a strong **electrostatic attraction** between the **oppositely charged ions** which are arranged in a **giant structure** of ions (**ionic lattice**). The lattice of sodium chloride is shown in four types of diagram.

1 Dot-and-cross

 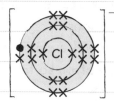

This diagram:

👍 shows why sodium atoms form 1+ ions and chlorine atoms form 1− ions

👎 does not show how the ions are arranged.

> You need to be familiar with the structure of sodium chloride but not other ionic compounds.

3 Ball and stick

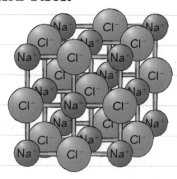

This diagram:

👍 shows the giant structure of ions

👍 shows how the electrostatic forces of attraction act in all directions (**ionic bonding**)

👎 suggests that the ionic bonds link specific ions, when the electrostatic attraction really acts in all directions

👎 does not show the electronic structure of the ions.

Properties of ionic compounds

1 high melting points: large amount of energy to break strong bonds

2 high boiling points: large amount of energy to break strong bonds

3 conduct electricity only when melted or dissolved in water – due to the ions not being free to move in the solid.

2 Two-dimensional

This diagram:

👍 shows how the ions are regularly arranged

👎 does not show how the ions are arranged in a giant structure

👎 does not show the electronic structure of the ions.

4 Three-dimensional

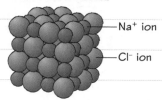

— Na⁺ ion
— Cl⁻ ion

This diagram:

👍 shows the giant structure of ions

👎 does not show the electronic structure of the ions.

Worked example

The structure of a compound is shown.

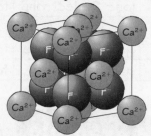

(a) Name the compound and the type of structure. **(2 marks)**

Calcium fluoride has a giant ionic lattice.

(b) This solid does not conduct electricity. What can be done so that the substance conducts electricity? **(2 marks)**

Melt the solid or dissolve the solid in water.

> The ions cannot move in the solid compound. How can the ions be freed from the lattice?

Covalent bonding

Covalent bonds are one of the three types of strong chemical bonds and form when non-metal atoms combine. A covalent bond is formed when atoms **share a pair of electrons**. Covalent bonds are found in **small molecules** (this and next page), **polymers** (page 21) and **giant covalent structures** (page 22).

Elements with small molecules

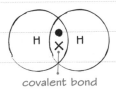
covalent bond

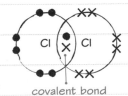
covalent bond

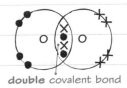
double covalent bond

The dot-and-cross diagrams show outer shell electrons only. Each atom in the molecules has a full outer shell. Pairs of electrons not used in bonding are called **lone pairs**.

Compounds with small molecules

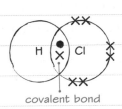
covalent bond

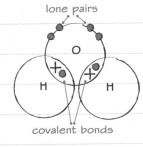
lone pairs
covalent bonds

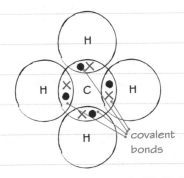
covalent bonds

Worked example

(a) Draw a dot-and-cross diagram of a molecule of nitrogen. **(2 marks)**

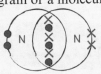

Nitrogen is in Group 5, so each atom has 5 outer electrons and will share 3 of them to give a full outer shell of 8 electrons.

(b) What type of bond is found in a nitrogen molecule? **(2 marks)**

a triple covalent bond

(c) Give the formula of a nitrogen molecule. **(1 mark)**

N_2

Now try this

1 Draw a dot-and-cross diagram to show the bonding in ammonia, NH_3. **(3 marks)**

2 Carbon and chlorine combine to form tetrachloromethane, CCl_4. This is a simple molecular compound with a similar structure to methane, CH_4.

(a) How would the bonds in tetrachloromethane form? **(3 marks)**

(b) Draw a dot-and-cross diagram to show the bonding in tetrachloromethane. **(3 marks)**

Small molecules

Some elements and compounds with atoms that are joined by covalent bonds form small molecules.

Representing small molecules: ammonia, NH_3

1 Dot-and-cross with shells

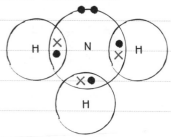

👍 shows the bonding electrons and lone pair

👎 does not show the shape

3 Dot-and-cross without shells

👍 shows the bonding electrons and lone pair

👎 does not show the shells

👎 does not show the shape

2 Two-dimensional

👍 shows the covalent bonds

👎 does not show the lone pair

4 Ball and stick

👍 shows how the atoms are arranged

👎 does not show the electrons

You should be able to draw diagrams for H_2, Cl_2, O_2, N_2, HCl, H_2O and CH_4 as well as ammonia. You can revise these molecules on the previous page.

Properties of substances with small molecules

- The covalent bonds between atoms in a molecule are **strong**.
- The **intermolecular forces** between molecules are **weak**.
- The intermolecular forces increase with the size of molecule.

These substances:

1 have a low melting point and boiling point

2 are usually gases or liquids

3 do not conduct electricity because they have no overall electric charge.

Worked example

Solid iodine, I_2, is heated until it forms a gas. Explain which particles are found in the gas. **(3 marks)**

The covalent bonds are not broken when the iodine melts and boils; only the weak intermolecular forces are broken, so iodine still exists as I_2 molecules in the gas state.

Now try this

The melting and boiling points of the halogens are shown.
(a) Give the state of these elements at room temperature. **(2 marks)**
(b) Explain the pattern in boiling points. **(3 marks)**

Consider the size of the halogen molecules.

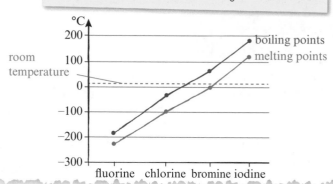

Polymer molecules

Some compounds with atoms that are joined by covalent bonds form very large molecules. Substances with these molecules, made up of **repeating units**, are called **polymers**.

A polymer molecule

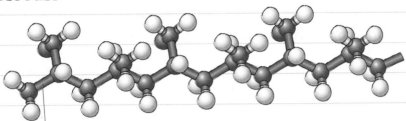

strong covalent bond   long-chain molecule

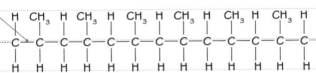

Properties of polymers

- Polymer molecules are very large.
- The intermolecular forces between the molecules are relatively strong compared with other covalent molecules.
- Polymers have relatively high melting points.
- Polymers are solid at room temperature.

Repeating units

The molecules above are made of lots of identical, repeating units in a chain. In this example, the repeating unit is:

$$-\overset{CH_3}{\underset{H}{C}}-\overset{H}{\underset{H}{C}}-$$

and the formula of the polymer is:

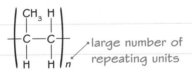

large number of repeating units

You can revise how polymers form on page 70.

Worked example

Part of a molecule is shown.

$$-C-C-C-C-C-C-$$

Explain why this covalent molecule has a melting point higher than most covalent molecules. **(2 marks)**

The molecule is a long chain, so there are relatively large intermolecular forces between molecules.

Now try this

The empirical formula is the simplest ratio of atoms of each element. How many hydrogen atoms are present for each carbon atom?

What is the empirical formula of the molecule shown in the Worked example? **(1 mark)**
A CH_2 B C_nH_{2n} C C_6H_{12} D CH

Diamond and graphite

Some covalently bonded substances have **giant covalent structures**, in which a huge number of atoms are joined in a lattice. These can be elements, such as carbon, or compounds, such as silicon dioxide.

Diamond

The carbon atoms in diamond form a lattice.

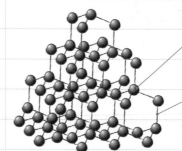

• each atom bonds to four others

• strong covalent bonds between atoms

Properties – diamond:
• is hard
• has a very high melting point
• does not conduct electricity.

Graphite

The carbon atoms in graphite form layers.

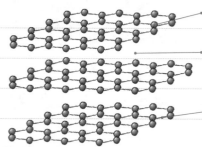

• each atom bonds to three others

• weak intermolecular forces between layers

• strong covalent bonds between atoms in a layer

Properties – graphite:
• is soft and slippery
• has a very high melting point
• conducts electricity in the same way as metals (see page 24).

Worked example

Diamond and graphite are both forms of carbon. Explain why graphite can conduct electricity but diamond cannot. **(4 marks)**

In graphite, each atom forms only three covalent bonds. One outer electron from each atom is delocalised and can carry charge from place to place. This allows graphite to conduct electricity. Diamond does not have delocalised electrons.

The weak intermolecular forces between the layers in graphite make it soft and slippery, whereas diamond is very hard.

The **delocalised** electrons allow graphite to conduct heat well, too. Graphite is similar to metals in having delocalised electrons (see page 24).

Silicon dioxide (silica)

The silicon and oxygen atoms in silica form a lattice:

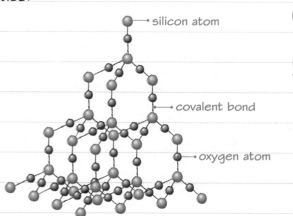

• silicon atom

• covalent bond

• oxygen atom

Now try this

(a) Explain why all substances with giant covalent structures have high melting points. **(2 marks)**

(b) Look at the structure of silica on the left. Which is silica's empirical formula? **(1 mark)**
   A   SiO
   B Si_2O
   C SiO_2
   D SiO_4

How many oxygen atoms are joined to each silicon atom? How many silicon atoms are joined to each of these oxygen atoms?

Graphene and fullerenes

Carbon exists in different structures. It can exist as the giant covalent structures diamond and graphite, which you can revise on the previous page. Carbon is also found as graphene, fullerenes and nanotubes.

Graphene

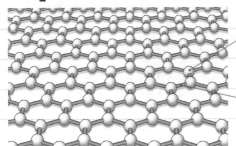

each carbon atom joined to three others

strong covalent bond

- Graphene is a single layer of graphite.
- The layer is one atom thick.

Uses of graphene:

 Graphene can be used to conduct electricity in electronic devices.

 Graphene can be used to make composites that are strong and of low density.

Fullerenes

Fullerenes:
- are molecules of carbon atoms
- have spherical or other hollow structures
- are based on hexagons of carbon
- may also have five-carbon and seven-carbon rings
- include Buckminsterfullerene (buckyballs), C_{60}, which has a spherical molecule.

Uses of fullerenes:

 in medicines to contain drugs

2 as lubricants

3 as a surface for a catalyst.

Cylindrical fullerenes: carbon nanotubes

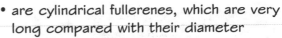

Carbon nanotubes:
- are molecules of carbon atoms
- are cylindrical fullerenes, which are very long compared with their diameter
- have high tensile strength
- are good conductors of electricity
- are good conductors of heat.

Uses of carbon nanotubes:

 in nanotechnology

 in electronics

 making new materials

 for reinforcing materials, such as in the frame of a tennis racket.

Worked example

(a) Describe the structure of graphene. **(2 marks)**

Each carbon is joined to three others by covalent bonds in hexagons, forming a layer one-atom thick.

(b) The holes in the graphene structure are large enough to allow water molecules through but not sodium ions or chloride ions. Suggest a potential use for graphene. **(1 mark)**

to filter seawater so that it can be used as drinking water

Now try this

Which properties of carbon nanotubes make them suitable for use in tennis rackets? **(2 marks)**

A tennis racket must not snap, but the material used to make it strong must not make the racket too heavy.

Metallic bonding

In metal elements and alloys, the outer electrons of the atoms are **delocalised** leaving a **lattice** of metal ions. The delocalised electrons completely cover the lattice.

Metallic bonding in sodium

positive sodium ion

The metal ions are arranged in a lattice.

The outer electrons of the atoms are lost, forming a 'sea' of electrons.

sodium ion (nucleus and inner electrons)

'sea' of delocalised electrons

The electrostatic attraction between the positive metal ions and the delocalised electrons is the strong metallic bond.

The regular arrangement of the sodium ions is called a lattice. The delocalised electrons are free to move around the whole lattice.

Metals and alloys

• Large amounts of metals, such as copper, gold, iron and aluminium, are used in everyday life.

• Pure metals are too soft for many uses, so other metals are mixed in to make alloys that are harder.

Worked example

Solder is a metal that is used to join electrical components. It must be melted so it can be applied. One solder is an alloy of tin and lead. Use the data in the table to answer the question.

Metal	Relative ability to conduct electricity	Melting point (°C)
tin	1.8	232
lead	1.0	327
solder	1.4	183

Explain why solder is a better choice for joining delicate electrical components than tin or lead on its own. **(3 marks)**

Solder melts at the lowest temperature. When the liquid solder is applied to the electrical components, it will not be too hot, so the components will not be damaged. Also, solder conducts electricity better than lead does.

Notice that solder, the alloy of tin and lead, has different properties from the metals it is made from.

Now try this

The melting points of sodium, magnesium and copper are given.

Metal	Melting point (°C)
sodium	98
magnesium	649
copper	1085

Remember that a cooking pan gets very hot.

(a) Explain why copper, not sodium or magnesium, is used to make a cooking pan. **(3 marks)**

(b) State why copper is used to make electrical wires. **(1 mark)**

Giant metallic structures and alloys

Metals have giant structures of ions with strong metallic bonding. You can revise the formation of metallic bonds on the previous page.

Properties of metals

- Metallic bonding is strong. Lots of energy is needed to break these bonds, so the melting points and boiling points of most metals are high.
- Most metals have high melting points, but one metal, mercury, is a liquid at room temperature and some melt easily.

You can revise the differences between metal and non-metal elements on page 9.

Metals are good conductors of electricity and heat

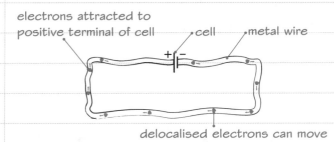

electrons attracted to positive terminal of cell ·cell ·metal wire

delocalised electrons can move

1 The metal can conduct electricity because the delocalised electrons are free to move and form a current.

2 When a metal is heated the delocalised electrons move and carry the heat energy, so metals are good conductors of heat.

Worked example

Pure copper contains layers of copper atoms. Copper is mixed with zinc to make brass.

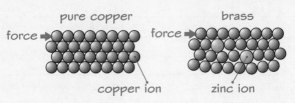

pure copper

force ➡

copper ion

brass

force ➡

zinc ion

(a) Explain why copper can be bent and shaped. **(1 mark)**

When a force is applied to the copper, the layers of ions can slide over each other, allowing a new shape to form.

(b) Explain why brass is harder than copper. **(2 marks)**

The larger zinc ions distort the layers of ions. The layers in brass cannot slide as easily as in copper.

Now try this

1 The table shows some information about iron and some types of steel (alloys of iron and carbon).
Explain why adding carbon to iron changes the strength of the metal.

Percentage carbon	0	0.2	0.4	0.8
Relative strength	1.0	1.8	2.6	3.3

(3 marks)

2 (a) Describe the structure and bonding in sodium. **(3 marks)**
   (b) Explain why sodium can conduct heat and electricity. **(2 marks)**
   (c) Explain why sodium can be bent into shape without shattering. **(2 marks)**

The three states of matter

Matter can be a solid, liquid or gas. Substances can change state when heated or cooled.

Changes of state

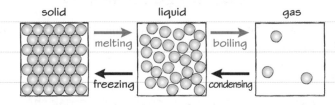

Particle theory

1 When a solid substance is heated the particles gain energy.

2 When the particles have sufficient energy to overcome the forces between them the substance melts.

3 If further heated, the substance boils when the particles have enough energy to overcome the forces between them.

4 The stronger the forces between the particles, the higher the melting and boiling points.

In the model shown the particles are represented by small, solid spheres. The model explains changes of state.

Worked example

Complete the table showing the type of particles and the forces present in the structure of each substance.

(5 marks)

Substance	Particles in structure	Forces present
metal or alloy	metal ions	strong metallic bonds between ions and delocalised electrons
noble gas	atoms	weak forces between atoms
other non-metal element or compound with low melting point	molecules	weak intermolecular forces between molecules
non-metal element or compound with high melting point	atoms	strong covalent bonds
ionic compound	metal ions and non-metal ions	strong ionic bonds between oppositely charged ions

---

Now try this

The melting points and boiling points of some carbon-containing substances are given.

Substance	Melting point (°C)	Boiling point (°C)
carbon dioxide	−56	−79
diamond	3527	4027
ethanol	−114	78
sodium carbonate	851	1600

 (a) Give the state of each substance at 25 °C.
(1 mark)

 (b) Explain the order of the melting points of the substances. **(5 marks)**

- Carbon dioxide is made of molecules.
- Diamond has a giant covalent structure.
- Ethanol is made of molecules.
- Sodium carbonate has a giant ionic lattice.

Nanoscience

Nanoscience relates to structures with a few hundred atoms, **nanoparticles**. Atoms and nanoparticles have properties different to the same material in bulk because of their high **surface area to volume ratio**.

Size of particles

1 Atoms: diameter of about 0.2 nm

2 Nanoparticles: structures of 1–100 nm, a few hundred atoms

3 Fine particles (PM$_{2.5}$): particles of diameter 100–2500 nm

4 Coarse particles (PM$_{10}$): particles of diameter 2500–10000 nm, **dust**

> **Maths skills** One nanometre (1 nm) is one millionth of a millimetre (10^{-9} m).

Maths skills: Surface area to volume ratio

1 This cube is 10 × 10 × 10

surface area of one face = 10 × 10 = 100

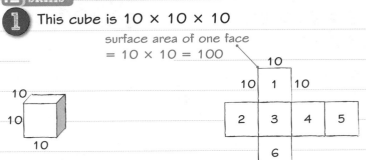

volume = 10 × 10 × 10 = 1000   surface area = 600

Surface area/volume ratio = 600/1000 = 0.6

2 This cube is 10 times smaller: 1 × 1 × 1

surface area of one face = 1 × 1 = 1

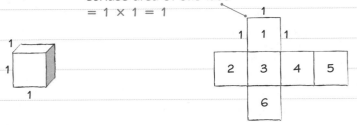

volume = 1 × 1 × 1 = 1   surface area = 6

Surface area/volume ratio = 6/1 = 6

As the cube got 10 times smaller, the surface area/volume ratio became 10 times larger.

Uses of nanoparticles

- in medicine to deliver drugs
- in medicine as synthetic skin
- in electronics
- in cosmetics
- in suncreams
- as catalysts in fuel cells
- in deodorants

The large surface area is one reason why nanoparticles can be used as catalysts.

New applications for nanoparticulate materials are being researched.

Smaller quantities of the material, in the form of nanoparticles, are needed to be effective.

Worked example

Nanoparticles are being used in suncreams. Evaluate their use. **(4 marks)**

Suncream that contains nanoparticles covers the skin better and so protects the body from the sun's UV rays better. Nanoparticles' high surface area to volume ratio means that more skin can be covered with the same amount of suncream. However, it is thought that, if nanoparticles penetrate the skin, they might cause damage to the DNA in living cells. If the suncreams are washed into the environment they could cause damage to marine life.

Now try this

1 Buckyballs are nanoparticles that consist of carbon atoms. Suggest how buckyballs can be used to get medicines into the body. **(2 marks)**

2 Vanadium dioxide is a blue powder. Scientists have developed a product that contains vanadium dioxide nanoparticles. These nanoparticles are invisible when they are applied to glass, but they significantly reduce the amount of heat energy transferred through the glass. Suggest **two** advantages of using this new coating when putting windows into a new building. **(2 marks)**

Extended response – Bonding and structure

You can revise the information for this question, which is about **bonding** and **structure**, on pages 22 and 23.

Worked example

Carbon exists in different forms. The diagrams of part of the structure of three of these forms are shown.

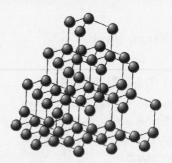 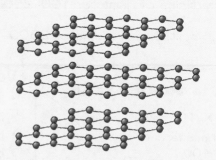 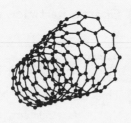

Explain the differences in the electrical conductivity of these three forms of carbon. **(6 marks)**

Diamond does not conduct electricity. Graphite and carbon nanotubes are good conductors of electricity.

Diamond exists as a giant covalent structure. Carbon is in Group 4, so each carbon atom has four outer electrons. In the diamond lattice, each carbon atom is joined by a covalent bond to **four** others. So, all of carbon's outer electrons are used in bonding. There are **no free electrons**, so diamond does not conduct.

Graphite consists of layers of carbon atoms. Each carbon atom is joined by a covalent bond to **three** others. So, each carbon atom has one more outer electron. These electrons are **delocalised**, and form a sea of electrons between the layers. The delocalised electrons can move and allow graphite to conduct electricity.

Carbon nanotubes consist of a layer of carbon atoms rolled into a cylinder. Similar to graphite, each carbon atom is joined to **three** others, leaving one extra that is delocalised, and these delocalised electrons allow carbon nanotubes to conduct electricity.

Don't forget to state what the differences in electrical conductivity are. Then use the diagrams to help explain the difference.

This answer is good because it gives full reasons why diamond does not conduct.

The diagram has been used well to describe the structure of graphite.

An electrical current consists either of a flow of electrons (in metals and some forms of carbon) or of ions (in ionic liquids or solutions).

Now try this

The melting points of sodium, chlorine and sodium chloride are given.

Substance	Melting point (°C)
sodium, Na	98
chlorine, Cl_2	−102
sodium chloride, NaCl	801

Explain the differences in these melting points by referring to the structure and bonding of each substance. **(6 marks)**

You must include the following:

1 What particles are present? How are these particles joined?

2 How are the particles arranged in bulk?

3 What forces/bonds have to be overcome to melt the substance?

Then compare the energy required in each case to explain the different melting points. Remember that sodium has a metallic structure, chlorine is made of molecules and sodium chloride has an ionic structure.

Relative formula mass

Relative formula mass

The **relative formula mass** of a substance is the sum of atomic masses of all the atoms shown in its formula.

Relative formula mass has the symbol M_r and it has no units – it is just a number.

Molecules

The formula of oxygen gas is O_2. Each molecule contains two oxygen atoms.

(A_r of O = 16)

M_r of O_2 = 16 + 16 = 32

Worked example

(a) Calculate the relative formula mass, M_r, of water, H_2O. **(1 mark)**
   (A_r of H = 1, A_r of O = 16)

   M_r of H_2O = 1 + 1 + 16 = 18

(b) Calculate the relative formula mass of $CaCO_3$.
   (A_r of Ca = 40, A_r of C = 12, A_r of O = 16)

   M_r of $CaCO_3$ = 40 + 12 + 16 + 16 + 16 = 100

Relative atomic masses
(A_r values) will normally be given to you in an exam question. If you need to look them up, they are on the periodic table given in the exam. Use the larger number in each box.

16
O
oxygen
8

⊞ Maths skills The '2' outside the brackets means that there are 2 oxygen atoms and 2 hydrogen atoms. So in the formula $Mg(OH)_2$, there are:
- 1 × Mg   • 2 × O   • 2 × H

⊞ Maths skills You can also use multiplication to get the same answer. Work out the M_r of OH first, multiply that by 2, then add the A_r of Mg:

M_r of OH = 16 + 1 = 17

M_r of $Mg(OH)_2$ = 24 + (2 × 17) = 58

Worked example

Calculate the relative formula mass, M_r, of magnesium hydroxide, $Mg(OH)_2$.
(A_r of Mg = 24, A_r of O = 16, A_r of H = 1)

M_r of $Mg(OH)_2$ = 24 + 16 + 16 + 1 + 1
= 58

Now try this

Use these relative atomic masses to answer the questions:
H = 1, C = 12, N = 14, O = 16, Na = 23, Mg = 24, Al = 27, S = 32

1 Calculate the relative formula masses of the following compounds.
   (a) magnesium sulfate, $MgSO_4$ **(1 mark)**
   (b) sodium carbonate, Na_2CO_3 **(1 mark)**
   (c) ethane, C_2H_6 **(1 mark)**

2 Calculate the relative formula masses of the following compounds.
   (a) aluminium hydroxide, $Al(OH)_3$ **(1 mark)**
   (b) ammonium carbonate, $(NH_4)_2CO_3$ **(1 mark)**
   (c) aluminium sulfate, $Al_2(SO_4)_3$ **(1 mark)**

Balanced equations and masses

A balanced equation shows that the particles in the reactants are rearranged to make the products. No atoms are lost or made. Revise how to balance an equation on page 14.

In the reaction $H_2 + Cl_2 \rightarrow 2HCl$: one molecule of H_2 reacts with one molecule of Cl_2 to make two molecules of HCl.

Law of conservation of mass

No atoms are lost or made in a reaction, so:

relative formula masses of all the reactants = relative formula masses of all the products.

Worked example

Show using relative formula masses that the equation $H_2 + Cl_2 \rightarrow 2HCl$ is balanced. (relative atomic masses: H = 1, Cl = 35.5)

$M_r(H_2) = 2 \times 1 = 2$
$M_r(Cl_2) = 2 \times 35.5 = 71$
$M_r(HCl) = 1 + 35.5 = 36.5$
relative formula mass of reactants =
$2 + 71 = 73$
relative formula mass of products =
$2 \times 36.5 = 73$

> **Maths skills** Two molecules of HCl are formed so 2 x 36.5 is the correct value.

relative formula mass of reactants =
relative formula mass of products
The equation is balanced. **(4 marks)**

Although the law of conservation of mass states that the mass of the reactants = the mass of the products, one product, CO_2, escapes, so the mass in the flask reduces.

Worked example

Some calcium carbonate is added to dilute hydrochloric acid.
$CaCO_3(s) + 2HCl(aq) \rightarrow CaCl_2(aq) + H_2O(l) + CO_2(g)$
The mass of the reaction container is monitored.

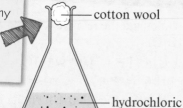

The cotton wool stops any acid spray escaping, so that only carbon dioxide escapes from the flask.

cotton wool

calcium carbonate — hydrochloric acid

balance

50 g of calcium carbonate is used.
(relative atomic masses: C = 12, O = 16, Ca = 40)
(a) Explain why the mass of the flask drops. **(1 mark)**
The carbon dioxide leaves the flask.
(b) Calculate the loss in mass. **(3 marks)**
Step 1: $M_r(CaCO_3) = 40 + 12 + (3 \times 16) = 100$
Step 2: $M_r(CO_2) = 12 + 16 \times 2 = 44$

> **Maths skills** The balanced equation shows $CaCO_3$ to CO_2 is in the ratio 1:1.

Step 3: 100 g of $CaCO_3$ would form 44 g CO_2
Step 4: So, 50 g of $CaCO_3$ would form 22 g CO_2
The loss in mass is 22 g.

Now try this

In these examples, state whether the recorded mass: A goes up; B goes down; C stays the same.
(a) Some copper is weighed, heated strongly, then re-weighed. **(1 mark)**
(b) Samples of sodium hydroxide and hydrochloric acid are weighed, mixed when neutralisation occurs, then the mixture is re-weighed. **(1 mark)**

(c) Some calcium carbonate is weighed. It is strongly heated when the following reaction occurs: $CaCO_3 \rightarrow CaO + CO_2$. The remaining solid is re-weighed. **(1 mark)**

Concentration of a solution

The concentration of a solution is a measure of the mass of dissolved substance (**solute**) in $1\,dm^3$ of solution.

Maths skills — Units

- ✓ A solution can be made by dissolving a solute in water (the **solvent**).
- ✓ The mass of the solute in grams (g) is measured.
- ✓ The volume of the solution is measured in dm^3.
- ✓ $1\,dm^3 = 1000\,cm^3$ (this volume is also known as one litre).

Concentrations in g/dm³

$$\text{concentration} = \frac{\text{mass (g)}}{\text{volume (dm}^3)}$$

The units of concentration are g/dm^3.

Maths skills The concentration of the acid is $4.00\,g/dm^3$. So, in the equation mass = concentration × volume, the volume must be in dm^3; $25\,cm^3$ is converted to dm^3 by dividing by 1000.

Worked example

(a) Calculate the mass of HCl in $25\,cm^3$ of $4.00\,g/dm^3$ of hydrochloric acid. **(3 marks)**

mass = concentration × volume

     = 4.00 × 25/1000

     = 0.1 g

(b) Calculate the mass of HCl that must be dissolved in water to make $20\,dm^3$ of this acid. **(3 marks)**

mass = concentration × volume

     = 4.00 × 20

     = 80 g

Worked example

Seawater contains approximately $35\,g/dm^3$ of sodium chloride. Calculate the mass of sodium chloride in $500\,dm^3$ of seawater. **(2 marks)**

mass = concentration × volume

     = 35 × 500

     = 17 500 g

Now try this

1 Calculate the concentration, in g/dm^3, of a solution where $9.8\,g$ of H_2SO_4 is dissolved in $500\,cm^3$ of water. **(3 marks)**

2 Which of the following has the largest mass of NaOH? **(1 mark)**

   A $2\,dm^3$ of $20\,g/dm^3$ NaOH solution

   B $40\,g$ of NaOH

   C $10\,dm^3$ of $10\,g/dm^3$ NaOH solution

Maths skills Convert the volume into dm^3 by using $1000\,cm^3 = 1\,dm^3$.

31

Reaction yields

The amount of a product obtained in a chemical process is called its **yield**.

Maximum theoretical yield

In a chemical reaction:
- no atoms are gained
- no atoms are lost.

For a given mass of reactants, it is possible to make only a certain maximum mass of products. This is the **theoretical yield**.

> The theoretical yield can be calculated using the balanced symbol equation for the reaction, and the relative masses of the substances involved.
> The actual yield is the amount obtained when the reaction is carried out and any purification steps have been done.

Actual yield

The **actual yield** is usually less than the theoretical yield.

REASONS FOR NOT GETTING THE THEORETICAL YIELD

> The reaction may be reversible, so it may not go to completion.

> Some of the product may be lost while separating it from the reaction mixture.

> There may be other reactions going on, so the reactants may produce other products (by-products) as well.

Worked example

When copper(II) carbonate is heated, it decomposes to form copper(II) oxide and carbon dioxide. The table shows the results from four experiments.

Experiment	Theoretical yield (g)	Actual yield (g)
1	0.5	0.4
2	1.0	0.8
3	1.5	1.2
4	2.0	1.2

$$\text{percentage yield} = \frac{\text{actual yield}}{\text{theoretical yield}} \times 100$$

The percentage yield is the actual yield compared with the theoretical yield:
- 0% means that no expected product was obtained.
- 100% means that all of the expected product was obtained.

(a) Suggest **one** reason for the difference between the theoretical yield and actual yield. **(1 mark)**

Some of the copper(II) carbonate may not have decomposed to form copper(II) oxide.

(b) Explain which experiment gave an anomalous result. **(2 marks)**

Experiment 4 was anomalous because the actual yield did not follow the trend. To follow the pattern, you would expect 1.6 g, but it was much less than this.

(c) Calculate the percentage yield for experiment 2. **(2 marks)**

Percentage yield = (0.8/1.0) × 100
= 80%

Now try this

1 (a) How is the actual yield usually different from the theoretical yield? **(1 mark)**

(b) Give **two** possible reasons for this difference. **(2 marks)**

2 Magnesium oxide can be made by reacting magnesium with oxygen:

$2Mg + O_2 \rightarrow 2MgO$

It is calculated that 1.2 g of magnesium oxide should be formed in an experiment. The actual yield was 0.9 g. Calculate the percentage yield. **(2 marks)**

Atom economy

The atom economy (or **atom utilisation**) tells us the percentage of the mass of the reactants that ends up as useful products in a reaction. The higher the atom economy, the less the waste.

 Maths skills   **Calculating atom economy**

The percentage atom economy is calculated using the masses in the balanced equation.

percentage atom economy = (mass of the desired product in the equation/mass of all the reactants in the equation) × 100%

Worked example

Bromoethane, C_2H_5Br, can be made from bromine and ethane.

$$C_2H_6 + Br_2 \rightarrow C_2H_5Br + HBr$$

% atom economy =   × 100

= (109/190) × 100

= 57%

HBr is formed in this reaction as a waste product. If a use can be found for HBr, then the atom economy for this reaction pathway would increase from 57% to 100%.

This worked example is about choosing a reaction pathway using atom economy.

Worked example

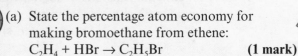

(a) State the percentage atom economy for making bromoethane from ethene:
$C_2H_4 + HBr \rightarrow C_2H_5Br$ **(1 mark)**

percentage atom economy = 100%

(b) Suggest why this method of making bromoethane might be preferred over the method using ethane.   **(2 marks)**

The atom economy is higher, so there is less waste. This preserves resources and removes the need to dispose of the waste products.

If there is only one product, **all** of the reactant atoms have gone into the product, so the atom economy is 100%.

1 A higher atom economy reaction is likely to be more **sustainable** – because less is wasted, raw materials will be preserved.

2 A higher atom economy reaction is likely to be more **economical** – because less is wasted, fewer raw materials need to be bought, and there will be less cost in the disposal of waste.

Reactions with one product

When ethene is polymerised many ethene molecules link together. The only product is poly(ethene).

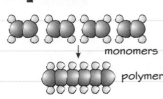
monomers
polymer

You can revise this on page 70.

Reactions like this have atom economies of 100%.

No atoms in the reactants are wasted.

Now try this

Calculate the percentage atom economy for producing iron by these two methods.
(relative atomic masses: C = 12, O = 16, Al = 27, Fe = 56)

(a) $Fe_2O_3 + 2Al \rightarrow 2Fe + Al_2O_3$ **(2 marks)**

(b) $Fe_2O_3 + 3CO \rightarrow 2Fe + 3CO_2$ **(2 marks)**

(c) Suggest one reason why method (b) is used in industry, even though the atom economy is lower.   **(2 marks)**

How is aluminium extracted from its ore?

Reactivity series

When metals react, their atoms form positive ions. The more easily the atoms form positive ions, the more reactive the metal. Metals can be arranged in order of their reactivity in the **reactivity series**.

The reactivity series

This can be written using the reactions with water and dilute acids.

	reaction with water	reaction with acid	
potassium	vigorous fizzing	very vigorous fizzing	
sodium	fast fizzing	very vigorous fizzing	
lithium	steady fizzing	very vigorous fizzing	
calcium	fizzing	vigorous fizzing	
magnesium	very slow fizzing	fast fizzing	increasing reactivity
carbon	–	–	
zinc	no observable reaction	fizzing	
iron	no observable reaction	very slow fizzing	
hydrogen	–	–	
copper	no observable reaction	no observable reaction	

These non-metals are often included in the reactivity series.

Reaction of metals with water and dilute acid

1 Metals react with water to form a metal hydroxide and hydrogen gas: e.g.
lithium + water → lithium hydroxide + hydrogen

2 Metals react with dilute acids to form a metal salt and hydrogen: e.g.
magnesium + hydrochloric acid →
magnesium chloride + hydrogen

Displacement reactions

1 A more reactive metal will displace a less reactive metal from its compound: e.g.
zinc + copper sulfate → zinc sulfate + copper

2 Metals above hydrogen in the reactivity series displace hydrogen from acids, for example:
magnesium + sulfuric acid →
magnesium sulfate + hydrogen

3 Carbon can displace metals below itself in the reactivity series from their compounds, for example:
carbon + zinc oxide → zinc + carbon dioxide

Worked example

1 (a) Compare the reactions of sodium and magnesium with cold water. **(3 marks)**

The metals both fizz making hydrogen, but sodium fizzes faster. Sodium moves around on the surface of the water.

(b) Give the reason for the difference in these reactions. **(2 marks)**

The metals form positive ions when they react. Sodium forms positive ions more easily, so it is more reactive.

Now try this

Metals **A**, **B** and **C** react with water. **A** and **B** fizz when added to water, but no observable reaction occurs with **C**, which reacts very slowly. When **A** is added to the chloride of **B**, no reaction occurs. When **B** is added to the chloride of **A**, a reaction occurs.
State and give evidence for the order of reactivity of **A**, **B** and **C**. **(3 marks)**

Remember that a more reactive metal can displace a less reactive metal from its compound.

Oxidation, reduction and the extraction of metals

Reactions in which substances gain oxygen atoms are **oxidation** reactions. Reactions in which substances lose oxygen atoms are **reduction** reactions.

Metals reacting with oxygen

- Many metals react with oxygen from the air.
- The metals **gain oxygen** to form metal oxides.
- An example of an oxidation reaction.
  $2Mg + O_2 \rightarrow 2MgO$
- Some metals, such as the alkali metals, react when left in air.
- Other metals react very slowly if left in air, but more quickly if heated.

Worked example

Explain whether these reactions are oxidation or reduction. **(4 marks)**

(a) Copper is heated in air to make copper oxide.

(b) Mercury oxide is heated and forms mercury and oxygen.

(a) Oxygen is added to the copper, so this is oxidation.

(b) Mercury oxide has oxygen removed, so this is reduction.

When a metal oxide forms a metal, reduction has occurred.

Extraction of metals

Many metals are found in the Earth as **metal oxides** or other compounds. When extracting the metal from its oxide, the oxygen atoms are lost. The loss of oxygen atoms is **reduction**.

Metals that are more reactive than carbon cannot be extracted using carbon. Electrolysis is used instead, which you can revise on page 41.

Metals that are less reactive than carbon can be extracted from their oxides by heating with carbon.

Unreactive metals are found as the metal itself.

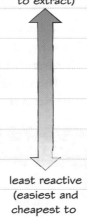

potassium		most reactive (most difficult and expensive to extract)
sodium	electrolysis of a molten compound	
lithium		
calcium		
magnesium		
aluminium		
carbon		
zinc	reduction from its oxide using carbon	
iron		
copper		
hydrogen		
silver	found as the element	least reactive (easiest and cheapest to extract)
gold		

Now try this

1 In the extraction of iron from iron ore, there are two reactions that occur:
   (a) Coke, a form of carbon, is heated and forms the gas carbon monoxide.
   (b) Iron oxide, the main compound in the iron ore, forms iron.
   Explain whether each reaction is oxidation or reduction. **(4 marks)**

2 Potassium is stored in oil. When a piece of potassium is removed from oil and cut, a silvery surface is seen. The silvery surface quickly goes dull.

   (a) Why is potassium stored in oil and not in air? **(2 marks)**

   (b) Explain the observations when potassium is cut. **(3 marks)**

The surface goes dull because the potassium is reacting and the product is not silvery.

35

Reactions of acids

Acids react with metals, **alkalis** and **bases** to make **salts**.

Reaction of metals with acids

metal + acid → metal salt + hydrogen

If the acid is **hydrochloric acid**, the metal salt will be a metal **chloride**.

If the acid is **sulfuric acid**, the metal salt will be a metal **sulfate**.

You should know how magnesium, zinc and iron react with acid, making the metal salts shown in this table.

	hydrochloric acid	sulfuric acid
magnesium	magnesium chloride	magnesium sulfate
zinc	zinc chloride	zinc sulfate
iron	iron chloride	iron sulfate

The reaction of dilute hydrochloric acid and metals to make a metal chloride

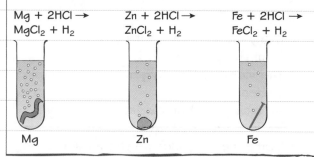

$Mg + 2HCl \rightarrow MgCl_2 + H_2$

$Zn + 2HCl \rightarrow ZnCl_2 + H_2$

$Fe + 2HCl \rightarrow FeCl_2 + H_2$

Mg   Zn   Fe

The order of the reactivity of the metals can be worked out using the speed of this reaction. You can revise reactivity on page 36.

The table shows the metal salt formed. The reactions all produce hydrogen gas.

Neutralisation reactions with acids

bases			metal carbonates
alkalis, soluble metal hydroxides	insoluble metal hydroxides	insoluble metal oxides	
→ salt + water			→ salt + water + carbon dioxide

If the reaction is with:

- **hydrochloric acid**, HCl, the salt is a metal **chloride**
- **sulfuric acid**, H_2SO_4, the salt is a metal **sulfate**
- **nitric acid**, HNO_3, the salt is a metal **nitrate**.

The ions in these acids, in addition to H^+, are Cl^-, SO_4^{2-} and NO_3^-.

Worked example

 (a) What is the formula for the salt potassium sulfate? **(1 mark)**

K_2SO_4

> As potassium is in Group 1 it forms a K^+ ion (see page 11), so two K^+ ions are needed to balance the 2− charge on the sulfate ion. (You can revise this on page 17.)

 (b) Write the balanced equation for the reaction between potassium carbonate and sulfuric acid. **(3 marks)**

$K_2CO_3 + H_2SO_4 \rightarrow K_2SO_4 + H_2O + CO_2$

 (c) What would you observe if potassium carbonate powder were added to dilute sulfuric acid? **(3 marks)**

The white powder disappears with fizzing, leaving a colourless solution.

Now try this

 Zinc reacts with hydrochloric acid to form zinc chloride, $ZnCl_2$.

(a) Write the balanced equation for the reaction. **(3 marks)**

(b) What would you **see** when this reaction occurs? **(2 marks)**

Core practical – Salt preparation

Core practical

Aim

To prepare a pure, dry sample of a salt.

Apparatus

- eye protection
- beaker
- measuring cylinder
- spatula and glass rod
- Bunsen burner
- tripod and gauze
- filter funnel and filter paper
- evaporating basin

Method

1. Measure out 50 cm³ of dilute sulfuric acid using a measuring cylinder, and pour into a beaker.

2. Place the beaker on a tripod and gauze and warm gently.

3. Add, using a spatula, a small amount of copper oxide and stir with the glass rod.

4. When all the copper oxide has reacted, add a further sample. Repeat until some unreacted copper oxide remains.

5. Filter the mixture into a beaker.

6. Heat the beaker gently using a **water bath** or electric heater until crystals begin to form.

7. Pour into an evaporating basin and leave in a warm place until crystals have formed.

8. Scrape the crystals on to some filter paper and pat dry.

Results

The equation for this reaction is:

$$CuO(s) + H_2SO_4(aq) \rightarrow CuSO_4(aq) + H_2O(l)$$

The black copper oxide forms a blue solution of copper sulfate. When the water is evaporated in Step 7, blue crystals are formed.

The preparation of a **soluble salt** can be carried out by reacting an acid with an **insoluble** substance: a metal, metal oxide, metal hydroxide or metal carbonate.

A **salt** is produced when an acid reacts with a metal, an alkali, a base or a metal carbonate. You can revise these reactions on page 36.

The type of salt depends on the acid used. Sulfuric acid forms a metal sulfate, hydrochloric acid produces a metal chloride, nitric acid makes a metal nitrate.

The acid is warmed so that the reaction happens faster.

The copper oxide is a black powder so, when some unreacted copper oxide is left, you will see a black solid in the beaker. This **excess** copper oxide is added to ensure that all of the acid has reacted.

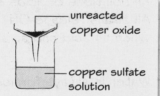

spatula
insoluble copper oxide
sulfuric acid

The mixture is filtered to remove the excess copper oxide.

unreacted copper oxide
copper sulfate solution

The solution is left so that the water evaporates slowly forming crystals.

copper sulfate crystals formed by evaporating the water

Worked example

Some copper nitrate crystals are prepared from copper carbonate.

(a) State the acid that you would use for this preparation.

nitric acid   **(1 mark)**

(b) Explain why an excess of copper carbonate is added to the acid.   **(2 marks)**

An excess of copper carbonate is added to ensure that all the acid has reacted. This means that, when the salt solution is heated to evaporate the water, an acidic solution is not left.

Now try this

Some copper nitrate crystals are prepared by the following method.

- Put some copper carbonate in a beaker and add acid until the solid has dissolved.
- Put the solution into an evaporating basin and heat until all of the water has evaporated.

Give **three** improvements to this method to make pure, dry, copper nitrate crystals.   **(3 marks)**

The pH scale

The pH scale, from 0 to 14, is a measure of how acidic or how alkaline a solution is.

A solution with pH less than 7 is **acidic**.

A solution with pH = 7 is **neutral**.

A solution with pH more than 7 is **alkaline**.

Acids produce H^+ ions in solution.

0 1 2 3 4 5 6 7 8 9 10 11 12 13 14

Solutions of **alkalis** contain OH^- ions.

more acidic ◄─────────────

solutions of increasing concentration of H^+

─────────────► more alkaline

solutions of increasing concentration of OH^-

Practical skills **Measuring pH**

1. Drops of **universal indicator** can be added to a solution. The colour in the mixture can be compared with the chart above and the pH read off.

2. A pH **probe** can be placed in the solution.

Worked example

(a) Name the salt produced in the reaction between potassium hydroxide solution and dilute nitric acid. **(1 mark)**

potassium nitrate

(b) Write the balanced equation, including state symbols, for the reaction. **(3 marks)**

$KOH(aq) + HNO_3(aq) \rightarrow KNO_3(aq) + H_2O(l)$

Reaction between the ions

- All solutions of acids contain H^+ ions.
- All alkalis release OH^- ions when dissolved in water.
- When an acid and an alkali are mixed, a **neutralisation** reaction occurs:

$$H^+(aq) + OH^-(aq) \rightarrow H_2O(l)$$

- Water is always formed when acids react with alkalis.

Worked example

Discuss the advantages and disadvantages of using universal indicator solution and a pH probe. **(4 marks)**

Universal indicator solution is easy to use. It has a range of colours so the pH chart can be used to find the pH. It is quite difficult to tell the exact colour so pH values are only approximate.

A pH probe is more complicated to use but it can give accurate pH values to one or two decimal places.

Now try this

1. Describe what colour you would see when universal indicator is added to:
   (a) hydrochloric acid
   (b) sodium hydroxide
   (c) pure water **(3 marks)**

2. Give two advantages of using a pH probe compared to using universal indicator solution. **(2 marks)**

Core practical – Titration

Core practical

Aim

To find the volume of alkali required to exactly neutralise 25 cm³ dilute hydrochloric acid.

Method

1 Wear eye protection.
2 Rinse the burette and then fill it with sodium hydroxide solution, making sure that the jet is filled.
3 Rinse the pipette then pipette 25.0 cm³ of dilute hydrochloric acid into a conical flask.
4 Add a few drops of indicator to the flask.
5 Add alkali from the burette, swirling the flask.
6 When close to the **end point** add the alkali drop by drop until the indicator just changes colour.

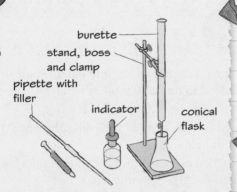

burette
stand, boss and clamp
pipette with filler
indicator
conical flask

Results

1 The volume reading on the burette is recorded at the start and at the end.
2 Repeat the experiment to get concordant results.

This method uses dilute hydrochloric acid, but other strong acids such as sulfuric acid and nitric acid can be used.

Risks:
• Acid and alkali solutions can be **corrosive** so eye protection must be worn.
• A safety filler is used.

The indicator must give a sharp colour change. One indicator is **phenolphthalein** which is pink in alkali and colourless in acid.

The burette and pipette are rinsed with the solutions that you are measuring in them to ensure that they are clean.

Maths skills

Sample results in cm³

End volume	24.70	26.20	25.80
Start volume	0.05	1.20	1.05
Titre	**24.65**	25.00	**24.75**

1 All volumes have 2 decimal places; the final one is always 0 or 5.

2 Subtract the start volume from the end volume to find the titre (volume used).

3 Choose the volumes that are all within 0.10 cm³ – the concordant results.

4 The result is the average of the concordant results:
= (24.65 + 24.75)/2 = 24.70 cm³

Worked example

Explain how the method used in the titration leads to accurate results. **(4 marks)**

• The alkali is added drop by drop near the end point to get an accurate volume.
• A white tile is used under the flask to see the indicator colour change more clearly.
• The bottom of the **meniscus** is used to read the volumes of liquids in the pipette and burette.
• The burette volumes are read to the nearest 0.05 cm³.
• The experiment is repeated until there are concordant results.

Now try this

Explain why the following procedures are used in a titration:
(a) A pipette is used and not a measuring cylinder to measure out the acid. **(1 mark)**
(b) The indicator that is used is **not** universal indicator. **(2 marks)**

Consider the colour changes that you would get with universal indicator.

Extended response – Quantitative chemistry

You can revise the information for this question, which is about **titrations**, on page 39.

Worked example

An experiment is carried out on an impure sample of sodium oxide. Some of the sodium oxide sample is weighed and then water is carefully added to make $100\,cm^3$ of solution. The reaction that occurs is:

$$Na_2O(s) + H_2O(l) \rightarrow 2NaOH(aq)$$

The sodium hydroxide solution formed is to be titrated with dilute hydrochloric acid. The results will then be used to find the amount of pure sodium oxide.

Explain how the experiment can be carried out to obtain accurate results for the volume of the dilute hydrochloric acid required to neutralise the alkaline solution.   **(6 marks)**

Making the solution

This answer is made clearer by using headings.

There are two procedures here – making the solution and then titrating it, so make sure that you cover both.

Wear eye protection.

On a balance weigh the amount of impure sodium oxide in the sample.

Tip the solid into a beaker.

Add distilled water and stir with a glass rod until all of the sodium oxide has reacted.

Pour the solution into a flask and rinse the beaker into the flask.

Add water so that the volume of the solution is $100\,cm^3$.

Add experimental detail such as 'rinse the beaker' (which ensures that all of the sodium hydroxide solution goes into the flask) which makes the method accurate.

The titration

Fill a burette with dilute hydrochloric acid using a funnel, then remove the funnel.

Make sure that the jet is filled.

Read the volume of the acid off the burette.

Pipette $25\,cm^3$ of the sodium hydroxide solution into a flask and add a few drops of phenolphthalein indicator.

Carefully add acid, swirling the flask.

The colour change is pink to colourless so, when the colour is faint, the sodium hydroxide is almost neutralised.

Add the acid drop by drop once the pink colour has become very faint.

When the mixture is colourless, read the volume again.

Repeat the titration until concordant results are achieved.

Now try this

Ethanol can be produced by the fermentation of sugars (**A**), which makes a dilute solution of ethanol, or the hydration of ethene (**B**), which produces a high purity of ethanol.

The equations are:

A $C_6H_{12}O_6 + 6O_2 \rightarrow 2C_2H_5OH + 2CO_2$

B $C_2H_4 + H_2O \rightarrow C_2H_5OH$

Calculate the atom economy of these two methods and explain the advantages of a reaction with a higher atom economy.

(6 marks)

Electrolysis

A melted or dissolved ionic compound will conduct electricity. **Electrolysis** is the decomposition of an ionic compound, when melted or dissolved, using electricity.

Ionic compounds conducting electricity

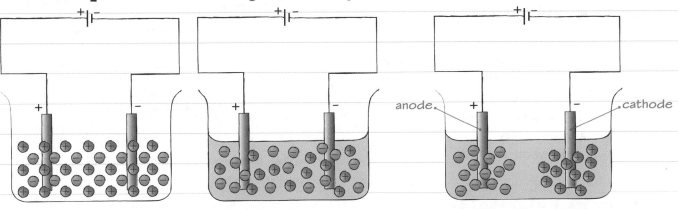

- In the solid, the ions are held firmly in a lattice.
- The ions cannot move so solid ionic compounds do not conduct electricity.
- You can revise giant ionic lattices on page 18.

- When melted or dissolved in water, the ions are released from the lattice.
- The ions can now move.
- Melted ionic compounds and dissolved ionic compounds conduct electricity.

- As electrolysis occurs: the positive ions move to the negative electrode (**cathode**). The negative ions move to the positive electrode (**anode**).
- Ionic compounds conduct electricity because the ions move and carry the current.

Oxidation and reduction of molten lead bromide

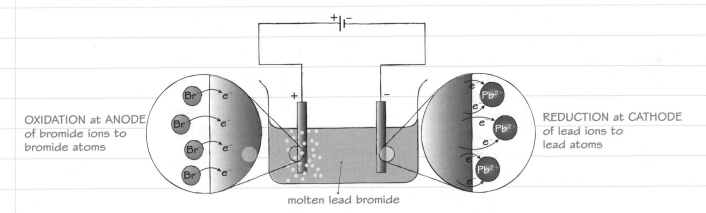

OXIDATION at ANODE of bromide ions to bromide atoms

REDUCTION at CATHODE of lead ions to lead atoms

molten lead bromide

Worked example

In the electrolysis of molten lead bromide:
(a) Explain why graphite is a suitable material to use for the electrodes. **(2 marks)**

Graphite conducts electricity and is inert.

(b) Explain why lead ions move to the cathode. **(2 marks)**

The lead ions are positive, so they are attracted to the negative cathode.

Now try this

Copper chloride, $CuCl_2$, can be electrolysed.
(a) Explain why copper chloride does not conduct electricity when solid but it does when liquid. **(2 marks)**

(b) Identify the products formed when molten copper chloride is electrolysed. **(2 marks)**

The copper chloride is molten, not dissolved, so no water is present.

Aluminium extraction

Aluminium is a widely used metal. It occurs naturally in the ore bauxite, which is mostly aluminium oxide. The aluminium is extracted from molten aluminium oxide by electrolysis.

Extraction of metals from their molten compounds

A molten mixture is made of aluminium oxide and cryolite, because this mixture has a lower melting point than pure aluminium oxide, saving energy costs.

Aluminium is widely used because:
- it has a low density
- it is strong when alloyed.

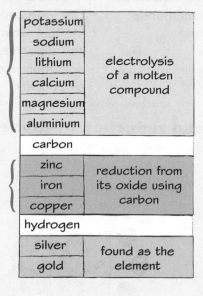

carbon anode   carbon cathode

The oxygen formed at the anode reacts with the anode forming carbon dioxide.

oxygen formed at anode

aluminium formed at cathode

molten mixture of aluminium oxide and **cryolite**

Why use electrolysis?

1 Electrolysis is expensive because of the cost of electricity – large amounts of energy are needed to melt the compounds and produce the electric current.

2 Carbon can displace metals (from their naturally occurring compounds) that are below carbon in the reactivity series (revise this on page 34). This is a much cheaper method.

3 Carbon will not react with the compounds of metals above carbon in the reactivity series, including aluminium, so electrolysis must be used.

4 Some metals react with carbon so electrolysis must be used.

Reactivity series

extracted by electrolysis	potassium		most reactive (most difficult and expensive to extract)
	sodium		
	lithium	electrolysis of a molten compound	
	calcium		
	magnesium		
	aluminium		
	carbon		
extracted by heating with carbon	zinc	reduction from its oxide using carbon	
	iron		
	copper		
	hydrogen		
	silver	found as the element	least reactive (easiest and cheapest to extract)
	gold		

Worked example

Explain why the carbon anodes in aluminium extraction have to be replaced regularly.   **(3 marks)**

Oxygen is formed at the anode. The anode is made of carbon, so it reacts with the oxygen forming carbon dioxide.

$$C + O_2 \rightarrow CO_2$$

Now try this

(a) Potassium can be obtained by the electrolysis of one of its salts. Explain what has to be done to the salt before electrolysis can occur.   **(1 mark)**

(b) Potassium can also be obtained by reacting potassium chloride with sodium. Write the word equation and the symbol equation for this reaction.   **(4 marks)**

Electrolysis of solutions

When a solution is electrolysed, hydrogen and oxygen may be produced at the electrodes; these come from the ions in the water making up the solution.

Ions from water

Water consists of molecules with covalent bonds, but a small proportion of water molecules break down to form hydrogen ions and hydroxide ions.

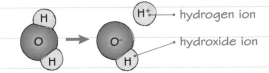
hydrogen ion
hydroxide ion

Reaction at the cathode

1 The cathode is the negative electrode, so the positive ion of the substance being electrolysed **and** H⁺ ions from the water are attracted to it.

2 If the metal in the substance being electrolysed is **more reactive** than hydrogen, then hydrogen is produced at the cathode.

3 If the metal in the substance being electrolysed is **less reactive** than hydrogen, then the metal is produced at the cathode.

The reactivity series on page 34 shows whether metals are more or less reactive than hydrogen.

Reaction at the anode

1 The anode is the positive electrode so the negative ion of the substance being electrolysed **and** OH⁻ ions from the water are attracted to it.

2 If the substance being electrolysed does not contain halide ions, then oxygen is produced at the anode.

3 If the substance being electrolysed contains halide ions (Cl^-, Br^-, I^-) then the halogen is produced at the anode.

You can revise the halogens on page 12.

Worked example

A solution of sodium chloride can be electrolysed.

(a) What product is formed at the anode? **(1 mark)**

chlorine

(b) What product is formed at the cathode? **(1 mark)**

hydrogen

(c) Which ions are left behind in the solution? **(2 marks)**

sodium ions and hydroxide ions

(d) Explain what you would see on adding litmus solution to the electrolyte after electrolysis. **(2 marks)**

The litmus goes blue because the mixture contains hydroxide ions, so it is alkaline.

Now try this

Predict the products that would form at the anode and at the cathode in the electrolysis of solutions of:

(a) potassium bromide   **(2 marks)**
(b) copper chloride   **(2 marks)**
(c) sodium sulfate.   **(2 marks)**

Sodium is above hydrogen in the reactivity series so hydrogen is formed.

A halide ion is present so the halogen is formed.

Practical skills

Core practical – Electrolysis

Core practical

Aim

To investigate what products are produced when different aqueous solutions are electrolysed.

Apparatus

- eye protection
- solutions of potassium bromide, calcium nitrate, copper nitrate, copper chloride
- graphite electrodes
- electrical circuit with d.c. supply
- electrolysis cell
- two small test tubes

Method

1   Set up the apparatus as shown.

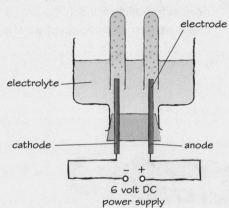

2   Pour one of the solutions (**electrolytes**) into the cell.
3   Turn on the d.c. supply.
4   Collect any gases produced in the test tubes and note observations at the electrodes.

> Aqueous solutions of ionic compounds conduct electricity because the ions are free to move.

> Graphite is inert. You can revise more about graphite on page 22.

> The positive ions move to the negative electrode (**cathode**).
> Either hydrogen (from the water) or the metal in the ionic compound is formed here.
> You can revise what is produced at the cathode on page 43.

> The negative ions move to the positive electrode (**anode**).
> Either oxygen (from the water) or a halogen (if the negative ion in the ionic compound is a halide ion) is formed here.
> You can revise what is produced at the anode on page 43.

> A **direct current** (d.c.) is needed for this experiment.

Results

	potassium bromide (aq)	calcium nitrate (aq)	copper nitrate (aq)	copper chloride (aq)
Observations at cathode	colourless gas *hydrogen*	colourless gas *hydrogen*	red–brown solid *copper*	red–brown solid *copper*
Observations at anode	orange solution *bromine*	colourless gas *oxygen*	colourless gas *oxygen*	pale-green gas *chlorine*

Now try this

 In the electrolysis of aqueous copper chloride:
(a) explain why the electrodes are made of graphite **(2 marks)**
(b) explain how you could test the pale-green gas to show that it is chlorine **(2 marks)**

 (c) explain why the process should be carried out in a fume cupboard **(1 mark)**
(d) suggest why a much smaller volume of chlorine is collected than expected. **(1 mark)**

Extended response – Chemical changes

You can revise the information for this question, which is about **extracting metals,** on pages 36, 37 and 42.

Worked example

The reactivity series of metals is shown, along with the elements carbon and hydrogen.
Suggest the method used to extract each of the metals from its ore, and explain why this method is used. **(6 marks)**

increasing reactivity ↑	sodium
	magnesium
	aluminium
	carbon
	zinc
	iron
	hydrogen
	copper
	gold

Copper and gold are unreactive. They are found uncombined in the Earth. This means that they are found as the element, so they do not need to be extracted from a compound.

Zinc and iron have medium reactivity. They can be found as oxides in the Earth. They can be extracted by heating the oxides with carbon. The metals can be reduced by carbon because they are below carbon in the reactivity series.

Aluminium, magnesium and sodium are more reactive than carbon so cannot be extracted from their compounds using carbon. Instead, their molten salts are electrolysed to produce the metal.

Some metals are found as elements. Those that are found as compounds can be extracted by heating with carbon or by electrolysis, so it makes sense to give an answer in three sections.

Use scientific terms such as **reduction** in your extended answers. This answer has used the reactivity series given to give a good explanation of the extraction method. Details of the extraction process are not needed.

The answer explains why the more expensive method **must** be used. You need to know details of aluminium extraction, but they are not needed for this answer.

Now try this

You are given samples of calcium, copper, gold, iron, magnesium, potassium, sodium and zinc. Design an experiment using the reactions of calcium, iron, magnesium and zinc with dilute hydrochloric acid so that you could use the results to put these metals into a reactivity series, and explain how you could adapt the experiment for copper, gold, potassium and sodium. **(6 marks)**

Remember to explain how you would control your variables.

45

Exothermic reactions

The total amount of energy in the universe at the end of a reaction is the same as that before the reaction happened. If the products have less energy than the reactants, then energy will be released to the surroundings. This is an **exothermic** reaction.

Reaction profile for an exothermic reaction

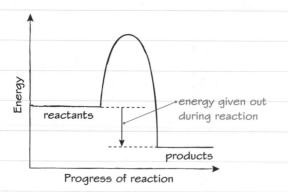

Progress of reaction

The reaction profile shows the energy change that occurs as the reaction happens.

Examples of exothermic reactions

1 combustion

2 some oxidation reactions

3 neutralisation

- The energy of the products is **lower** than that of the reactants in an exothermic reaction.
- Energy is transferred to the surroundings.
- The temperature of the surroundings increases.

Hand warmers

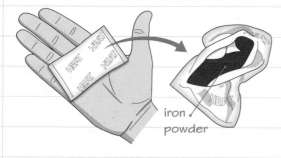

iron powder

$4Fe + 3O_2 \rightarrow 2Fe_2O_3$

1 When the hand warmer is activated, the iron oxidises.

2 The oxidation reaction is exothermic.

3 Heat is released to the surroundings (your hands).

Worked example

The diagram shows a self-heating drink.

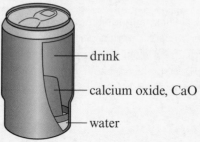

— drink

— calcium oxide, CaO

— water

Explain how the self-heating can works, including a balanced equation.   **(4 marks)**

When the can is activated, the calcium oxide reacts with the water.
$CaO + H_2O \rightarrow Ca(OH)_2$
The reaction is exothermic. Heat is released which heats up the drink.

Now try this

Some acid and alkali are mixed. The temperature was recorded before and after mixing, as shown. Calculate the temperature change and state whether the reaction is exothermic or endothermic.   **(2 marks)**

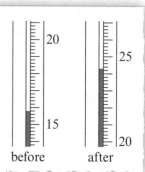

before   after

Endothermic reactions

The total amount of energy in the universe at the end of a reaction is the same as that before the reaction happens. If the products have more energy than the reactants, then energy must be absorbed from the surroundings. This is an **endothermic** reaction.

Reaction profile for an endothermic reaction

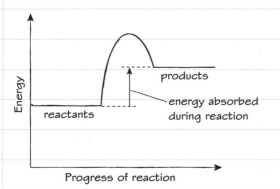

The reaction profile shows the energy change that occurs as the reaction happens.

Examples of endothermic reactions

1 thermal decomposition

2 reaction of citric acid with sodium hydrogen carbonate

- The energy of the products is **higher** than that of the reactants in an endothermic reaction.
- Energy is absorbed from the surroundings.
- The temperature of the surroundings decreases.

Some endothermic reactions

1 Calcium carbonate decomposes when heated.

$$CaCO_3 \rightarrow CaO + CO_2$$

2 Citric acid reacts with sodium hydrogen carbonate.

$$C_6H_8O_7 + 3NaHCO_3 \rightarrow$$

$$3CO_2 + 3H_2O + C_6H_5O_7Na_3$$

You do not need to learn this equation.

Worked example

The diagram shows a sports injury pack.

INSTANT COLD PACK
1. Squeeze pack   2. Shake
3. Apply to patient   4. Dispose of used pack

(a) Why is the pack squeeezed? **(1 mark)**

Squeezing breaks open the inner bag of water.

(b) Why does the pack get cold? **(2 marks)**

When the water is released the solid in the pack dissolves. This dissolving is endothermic.

(c) Why is the pack single-use only? **(1 mark)**

The dissolving cannot be reversed.

Now try this

(a) What would you observe when solid sodium hydrogen carbonate is added to a solution of citric acid? **(2 marks)**

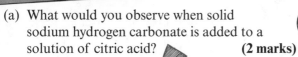

Use the equation above to help you.

(b) How would you show that this reaction was endothermic? **(4 marks)**

Core practical – Energy changes

When an acid reacts with a metal, a metal carbonate or an alkali, the **energy change of the reaction** will cause a change in the temperature of the mixture. Different concentrations of acid will give different temperature rises.

Core practical

Aim

To investigate the temperature change that occurs when different concentrations of acid react with zinc.

> This experiment uses zinc powder, but metal carbonates could also be used, or solutions of metal carbonates or alkalis.

Apparatus

- eye protection
- dilute acid
- zinc powder
- measuring cylinder
- glass beaker
- polystyrene cup
- thermometer

Method

1 Set up the apparatus as shown.

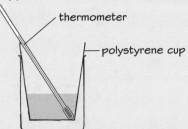

thermometer

polystyrene cup

> A polystyrene cup is used because polystyrene is a poor conductor of heat, so most of the heat energy released by the reaction is kept in the container.

2 Measure 50 cm³ of the dilute hydrochloric acid.

3 Add the acid to the polystyrene cup and record its temperature.

4 Add all of the zinc powder, which needs to be in excess.

5 Stir with the thermometer until the temperature stops rising.

6 Record the maximum temperature.

7 Repeat the experiment with acid–water mixtures:
40 cm³ acid + 10 cm³ water; 30 cm³ acid + 20 cm³ water;
20 cm³ acid + 30 cm³ water; 10 cm³ acid + 40 cm³ water.

> The zinc powder has to be stirred vigorously to ensure that it reacts fully.

> As this is an exothermic reaction, the **maximum** temperature is recorded.

> Most reactions cause a temperature increase: they are **exothermic**. Some cause a temperature decrease: they are **endothermic**. You can revise these types of reaction on pages 46 and 47.

Sample results

Volume of acid (cm³)	10	20	30	40	50
Temperature rise (°C)	5.1	10.0	14.9	20.1	25.0

> The total volume of the acid/water mixture is always the same so that the same volume of solution is heated up each time.

> The more acid present, the higher the temperature rise.

Now try this

> Remember that the zinc was already in excess.

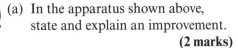

(a) In the apparatus shown above, state and explain an improvement. **(2 marks)**

(b) The experiment using 50 cm³ of acid is repeated, but using twice as much zinc powder. Estimate the temperature rise and explain your answer. **(3 marks)**

Activation energy

The **activation energy**, E_a, is the energy needed for a reaction to occur.

Exothermic reaction

(see page 46)

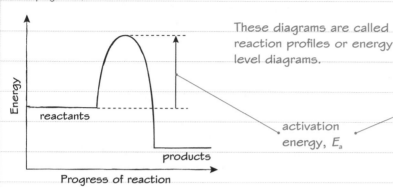

These diagrams are called reaction profiles or energy level diagrams.

activation energy, E_a

Endothermic reaction

(see page 47)

How reactions occur

• For a reaction to occur, particles of the reactants must collide.
• When the particles collide, they must have enough energy for a reaction to occur.
• The minimum energy needed for a successful collision (so that products form) is the activation energy.

You can revise factors affecting the rate of a reaction on pages 55 and 56.

Worked example

The energy level diagram represents an exothermic reaction.

(a) Explain what the energy shown as A represents.
(2 marks)

The activation energy, which means that this is the energy needed to start the reaction.

(b) Draw a line to represent the change in energy level when a catalyst is added. **(1 mark)**

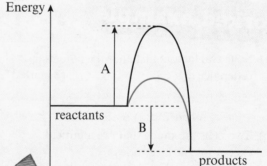

In this diagram, B represents the overall energy change. A catalyst provides an alternative pathway with a lower activation energy for the chemical reaction. (You can revise catalysts on page 58.)

Now try this

1 Explain how you would know that the energy level diagram in the Worked Example shows an exothermic reaction. **(2 marks)**
2 Hydrogen burns in oxygen to form one product only.
   (a) Write the equation for the reaction. **(2 marks)**
   (b) Explain why a flame is needed to start the reaction. **(1 mark)**
3 Referring to the Worked Example, explain why the line goes up then drops to a lower level. **(4 marks)**

Cells

A **cell** contains chemicals that react to produce electricity. Two or more cells can be connected together to form a **battery**, which has a higher voltage.

Chemical cell

A **chemical cell** can be made by connecting two metals that are in contact with an electrolyte.

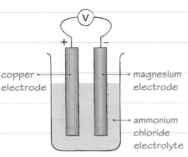

copper electrode · magnesium electrode · ammonium chloride electrolyte

Advantages and disadvantages

 portable

👍 rechargeable cells can be reused (they are recharged by applying an external current that reverses the chemical reactions in the cell)

👎 when one of the reactants in a non-rechargeable cell (such as in an **alkaline battery**) has been used up, the chemical reactions stop

👎 difficult to dispose of

Fuel cell

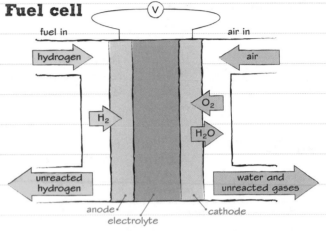

fuel in · hydrogen · air in · air · O_2 · H_2 · H_2O · unreacted hydrogen · water and unreacted gases · anode · electrolyte · cathode

cell reaction: $2H_2 + O_2 \rightarrow 2H_2O$

A **fuel cell** is continuously supplied with fuel (such as hydrogen) and oxygen (or air). The fuel is oxidised, giving a **potential difference**.

Advantages and disadvantages

👍 constant voltage

👍 does not run down if fuel is continuously supplied

👍 only waste product of hydrogen fuel cell is water

👎 not portable

Worked example

(a) State **two** factors that alter the potential difference in a cell. **(2 marks)**

1 the electrolyte

2 the metals used for the electrodes

(b) Two cells are made that are identical, except for the electrodes.
cell **A**: magnesium and copper
cell **B**: zinc and copper
Use the reactivity series to explain which cell will have the higher potential difference. **(2 marks)**

Cell A has the higher potential difference because the gap between magnesium and copper in the reactivity series is bigger than the gap between zinc and copper.

Worked example

Hydrogen fuel cells can be used in cars.
Give **one** advantage and **one** disadvantage of each of the following for use in a car.

(a) petrol **(2 marks)**

Advantage – car can travel long distances on one tank of fuel; Disadvantage – petrol is non-renewable fuel.

(b) hydrogen fuel cell **(2 marks)**

Advantage – hydrogen is a renewable fuel; Disadvantage – hydrogen has to be stored in heavy tanks.

(c) rechargeable battery **(2 marks)**

Advantage – if the electricity is generated not using fossil fuels, no carbon dioxide is emitted; Disadvantage – it takes time to charge battery when it runs down.

Now try this

Explain why using a hydrogen fuel cell might lead to global warming, even though it does not cause pollution when it is working. **(3 marks)**

The hydrogen will be made by the electrolysis of water. Think about how the electricity used in this electrolysis is generated.

Extended response – Energy changes

You can revise the information for this question, which is about **endothermic reactions**, on pages 47 and 48.

Worked example

The diagram shows a sports injury pack. When the pack is squeezed, a bag of water breaks and the solid in the pack dissolves in the water. This dissolving is an endothermic reaction.

You are given 10 g of two solids, potassium chloride and ammonium nitrate.

Design an experiment to find out which solid would be best to use in a sports injury pack. **(6 marks)**

- Place a polystyrene cup in a beaker.
- Measure 50 cm³ of water with a measuring cylinder.
- Pour the water into the polystyrene cup.
- Record the temperature of the water with a thermometer.
- Tip in one of the solids.
- Stir the mixture carefully.
- When the temperature stops decreasing, record the lowest temperature.
- Repeat the experiment with fresh water using the other solid.
- Wear safety glasses and gloves during the experiment.
- The solid that produces the biggest fall in temperature would be the best to use in a sports injury pack.

 A polystyrene cup is used to minimise heat being absorbed from the outside.

 The use of bullet points makes this answer clear, because it is describing a step-by-step method.

 A sports injury pack is designed to reduce temperature, so the solid that gives the biggest temperature drop would be best.

Other factors that might be important include:
- whether the solids are hazardous if the pack splits
- the cost of the solids.

Now try this

When dilute hydrochloric acid reacts with sodium hydroxide solution, the neutralisation is exothermic. Design an experiment to measure the temperature rise when 50.0 cm³ sodium hydroxide solution is completely neutralised. **(6 marks)**

 Add the hydrochloric acid bit by bit until you are sure that the sodium hydroxide is *completely* neutralised.

Rate of reaction

 Practical skills The **rate of reaction** can be found by measuring the quantity of reactant used or the quantity of product formed over time.

 Maths skills ## The rate of reaction

1 If the quantity of reactant is measured:

mean rate of reaction =
$$\frac{\text{quantity of reactant used}}{\text{time taken}}$$

2 If the quantity of product is measured:

mean rate of reaction =
$$\frac{\text{quantity of product formed}}{\text{time taken}}$$

As well as the time in seconds, the quantity of any reactant or product can be measured in an experiment. You could measure mass in grams or volume of a gas in cm^3. The units of rate of reaction in each case are:
- measuring mass: rate in g/s
- measuring volume: rate in cm^3/s

Measuring the rate of reaction of marble chips with dilute hydrochloric acid

The equation for the reaction is: $CaCO_3(s) + 2HCl(aq) \rightarrow CaCl_2(aq) + H_2O(l) + CO_2(g)$

This reaction is monitored by measuring the mass of the flask and contents. This falls as the carbon dioxide gas escapes.

The cotton wool allows CO_2 to escape but prevents any acid spray leaving.

The rate of reaction is calculated by working out the mass of product (carbon dioxide) formed over time.

Sample results

Time (min)	Mass (g)	Time (min)	Mass (g)
0	135.60	3	135.25
1	135.43	4	135.22
2	135.34	5	135.20

Worked example

Use the results of the experiment to calculate:
(a) The rate of reaction in the first minute. **(2 marks)**

(135.60 – 135.43)/60 = 0.0028 g/s
(b) The rate of reaction in the second minute. **(2 marks)**

(135.43 – 135.34)/60 = 0.0015 g/s
(c) The rate of reaction over the first 3 minutes. **(2 marks)**

(135.60 – 135.25)/180 = 0.0019 g/s

 Maths skills The time is calculated in seconds (1 minute = 60 seconds), giving the units g/s.

 ## Now try this

 (a) From the results of the experiment above, calculate the rate of reaction over 5 minutes. **(1 mark)**

(b) Explain why the data in the Worked example shows that the rate of reaction in the second minute is lower than the rate of reaction in the first minute. **(1 mark)**

Rate of reaction on a graph

Practical skills A graph can be drawn of the quantity of product formed or quantity of reactant used up against time. The **gradient** of this graph is the rate of reaction.

Measuring the rate of reaction of marble chips with dilute hydrochloric acid

One method to monitor this reaction (by measuring the mass of product formed) was described on the previous page. You can also see the balanced equation on that page.

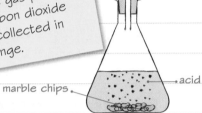

gas syringe

This method monitors the volume of the gas product, CO_2. The carbon dioxide given off is collected in the gas syringe.

marble chips → acid

Sample results

Time (min)	Volume (cm³)	Time (min)	Volume (cm³)
0	0	3	178
1	87	4	193
2	132	5	204

Maths skills Showing the results on a graph

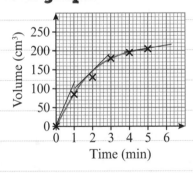

1 The quantity of product formed or reactant used up (on the y axis) against time (on the x axis) can be plotted on a graph.

2 Choose axes so that the data points occupy at least half of the graph paper.

3 Label the axes with the units:

y axis: volume (cm³)

x axis: time (min)

4 Take great care plotting the points.

5 Draw a best-fit straight line or smooth curve. **Do not** join one point to the next. Ignore **anomalous** points (such as the value at 2 minutes).

Anomalous points are ones that do not fit on a straight line or a smooth curve. You will normally be told in the exam whether a straight line or a smooth curve is needed.

How the rate changes

- On the volume–time graph, measuring the gradient of the tangent gives the rate of reaction.
- On the graph above, the slope of the tangent is steeper at 2 minutes than it is at 4 minutes.
- This shows that the rate of reaction falls as the reaction proceeds.

Now try this

Calculate the average rate of reaction over the first 2 minutes, in cm³/min.

(2 marks)

Collision theory

Collision theory explains how changes in conditions affect the rate of reaction. The concentration of reacting solutions, pressure of reacting gases, surface area of reacting solids, the temperature of the reaction and the presence of a catalyst will all alter the rate.

A mixture of particles undergoing a reaction

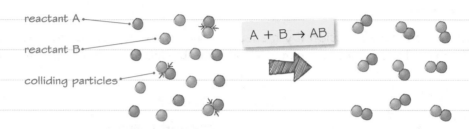

reactant A
reactant B
colliding particles

$A + B \rightarrow AB$

Only a small proportion of the collisions result in a reaction (see page 56).

Collision theory states that for a reaction to occur:

1 the reactant particles must collide

2 the reactant particles must have a minumum amount of energy (the **activation energy**).

(You can revise activation energy on page 49.)

How to increase the rate of reaction

1 The **frequency of collisions** between the reactant particles can be increased.

2 The energy of the reactant particles can be increased.

The ways of achieving this are considered on this and the next few pages.

Worked example

Explain the effect of doubling the concentration of the reactants on the rate of reaction.   **(3 marks)**

1 The number of reactant particles in the container are doubled.

2 The frequency of collisions will be doubled.

3 The rate of reaction will be doubled.

The collision theory predicts this increase of rate, and many, but not all, reactions match this prediction.

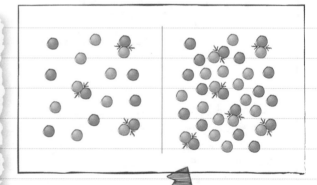

number of particles × 2
collision frequency × 2

Now try this

Magnesium ribbon reacts with dilute hydrochloric acid. Explain how increasing the concentration of the hydrochloric acid alters the rate of the reaction.
(3 marks)

Do not forget to mention the particles and the frequency of collisions in your answer.

Rate: pressure, surface area

The collision theory about rates of reaction is explained on the previous page. Altering the surface area of reacting solids and altering the pressure of reacting gases are two factors that will alter the rate of reaction.

Maths skills Surface area to volume ratio

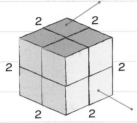
surface area of one face

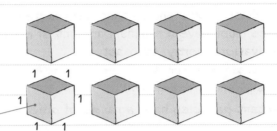
each cube has six faces

Single large cube

surface area = 6 × (2 × 2) = 24

volume = 2 × 2 × 2 = 8

surface area : volume ratio = 24/8 = 3

Eight small cubes

surface area = 8 × (6 × (1 × 1)) = 48

volume = 8 × (1 × 1 × 1) = 8

surface area : volume ratio = 48/8 = 6

✓ If a solid reactant is broken up into smaller pieces the surface area:volume ratio increases.

✓ The solid will have more exposed surface for other reactants to collide with.

✓ There will be more frequent collisons.

✓ The solid with smaller pieces will react faster.

Effect of pressure

The higher-pressure gas has closer together molecules.

The closer together the molecules in the container, the higher the frequency of collisions between reactant molecules.

The higher the frequency of collisions, the higher the rate of reaction.

Low pressure   High pressure

In many cases, if the pressure is doubled, the number of reactant molecules in an equal volume is doubled, so the collision frequency doubles and the rate of reaction doubles.

Worked example

Some marble chips are placed in an excess of dilute hydrochloric acid. The gas released is measured each minute. The results are plotted on the graph.

(a) Draw the graph that you would get if the experiment were repeated with an identical mass of marble chips, but with smaller chips, and label the line A. **(2 marks)**

(b) Draw the graph that you would get if the experiment were repeated with half the mass of very large marble chips, and label the line B. **(2 marks)**

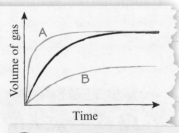

The reaction is slower, so the graph is less steep.

As there is half the mass of chips, only half the volume of carbon dioxide will be formed.

Now try this

With three times as many molecules, how much will the collision frequency increase by?

In the reaction $H_2 + I_2 \rightarrow 2HI$, an equal number of molecules of hydrogen and iodine gases are mixed at a high temperature. The time taken for the maximum amount of HI to form is measured.

Explain the effect on the time taken for the HI to form if three times as many molecules of each gas were placed in a container of the same volume at the same temperature. **(4 marks)**

55

Rate: temperature

Collision theory (described on page 54) states that increasing the frequency of collisions and increasing the energy of the reacting particles will increase the rate of a reaction.

Increasing the energy of reacting particles

The reaction profile shows that the reactant particles must have a minimum amount of energy, the activation energy, to react.

(You can revise reaction profiles on pages 46 and 47, and activation energy on page 49.)

- Reactant particles have a range of different energies.

- The higher the temperature, the higher the proportion of reactant particles with at least the activation energy.

- The higher the temperature, the higher the proportion of collisions that are 'successful' (lead to a reaction).

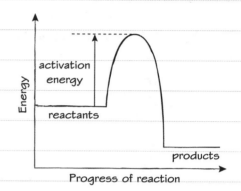

Reactant particles with more energy move faster, so the frequency of collisions is also increased at a higher temperature.

Worked example

Zinc reacts with dilute hydrochloric acid.

(a) Write the balanced equation for the reaction. **(3 marks)**

$Zn + 2HCl \rightarrow ZnCl_2 + H_2$

(b) Explain how you could use the apparatus shown to see whether an increase in temperature of the acid increases the rate of reaction. **(4 marks)**

- Add a **measured volume** of acid to the flask.

- Add a **known mass** of zinc to the acid.

- Seal the flask and measure the volume of gas given off every minute for a certain time.

- Repeat the experiment, heating the acid to a different temperature before adding it to the flask.

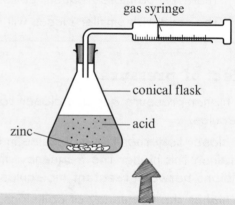

Practical skills
When carrying out this experiment, you would have to:
- heat the acid, wearing safety goggles (the acid may be corrosive)
- keep the experiment away from a naked flame (the hydrogen is flammable).

Now try this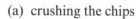

1 Marble chips react with hydrochloric acid:
$CaCO_3(s) + 2HCl(aq) \rightarrow$
$CaCl_2(aq) + H_2O(l) + CO_2(g)$
State and explain the effect of the following on the rate of the reaction:
(a) crushing the chips **(3 marks)**
(b) diluting the acid **(3 marks)**
(c) heating the acid. **(4 marks)**

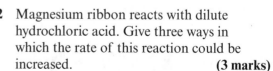 2 Magnesium ribbon reacts with dilute hydrochloric acid. Give three ways in which the rate of this reaction could be increased. **(3 marks)**

Practical skills

Core practical – Rate of reaction

The change in concentration of a reactant affects the rate of reaction.

Core practical

Aim

Sodium thiosulfate reacts with dilute hydrochloric acid. The aim is to monitor the effect of the change of concentration of the sodium thiosulfate solution on the rate of this reaction.

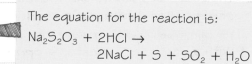
The equation for the reaction is:
$$Na_2S_2O_3 + 2HCl \rightarrow 2NaCl + S + SO_2 + H_2O$$

Apparatus

- eye protection
- dilute hydrochloric acid
- sodium thiosulfate solution
- measuring cylinders
- flask
- black cross on white paper
- timer

Method

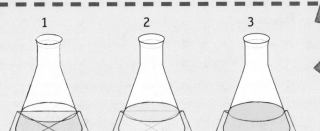

The two reactant solutions are colourless. The sulfur product forms as a yellow precipitate.

Sulfur dioxide causes breathing difficulties, so the experiment should be carried out in a well-ventilated room.

1. Measure 50 cm³ of sodium thiosulfate solution in a measuring cylinder and transfer into a flask. With a separate measuring cylinder, measure 10 cm³ of dilute hydrochloric acid. Add the hydrochloric acid to the flask, start timing and swirl the flask. Place the flask on to a black cross.
2. A yellow sulfur precipitate will form in the flask.
3. Record the time taken when the cross can no longer be seen.
4. Repeat the experiment with these mixtures of sodium thiosulfate solution and water:

sodium thiosulfate (cm³)	40	30	20	10
water (cm³)	10	20	30	40

The total volume of the acid/water mixture is always the same so that the same volume of solution is used each time.

Sample results

volume of sodium thiosulfate (cm³)	50	40	30	20	10
time (s)	21	33	48	69	173

The higher the concentration of the sodium thiosulfate solution, the faster the rate of reaction.

🖩 Maths skills — Uncertainty in results

Two students repeat the experiment.

Student 1's results with 50 cm³ of sodium thiosulfate are 20 s, 21 s and 22 s. The average is 21 s, with an uncertainty of ±1 s.

Student 2's results are 18 s, 21 s and 24 s. The average is 21 s with an uncertainty of ±3 s.

Now try this

Calcium carbonate is reacted with different concentrations of hydrochloric acid. The time taken for 250 cm³ of carbon dioxide to be released is measured.

(a) What pattern is shown in the results? **(1 mark)**
(b) Suggest the time that would be recorded with acid of relative concentration 6. **(2 marks)**

Relative concentration	1	2	4	8	16
Time (s)	500	305	200	113	45

Catalysts

Catalysts are substances that change the rate of reaction but are not used up in the reaction.

Catalysts and activation energy

- The diagram shows the reaction profile for an exothermic reaction.

  You can revise this on page 46.
- The red curve shows the reduced activation energy when a catalyst is used.

  You can revise activation energy on page 49.
- With a catalyst, a greater proportion of the reacting particles have the activation energy, so a greater proportion of the collisions will be 'successful'.

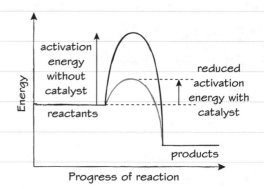

Properties of catalysts

- Catalysts provide an **alternative reaction pathway** with lower activation energy.
- Catalysts are specific – different reactions need different catalysts.
- In biological systems, catalysts are called **enzymes**.

  You can revise an example – the fermentation reaction – on page 68.

Worked example

Hydrogen peroxide decomposes to make water and oxygen:

$$2H_2O_2(aq) \rightarrow 2H_2O(l) + O_2(g)$$

> The catalyst does not appear in the chemical equation.

A catalyst for this reaction is manganese(IV) oxide, MnO_2. Identical volumes of hydrogen peroxide solution were allowed to decompose for 10 minutes. One sample was just hydrogen peroxide solution (with no added solid), one had added manganese(IV) oxide, one had added lead(IV) oxide and one had added sodium chloride. The volume of oxygen given off was measured.

Results

Solid added	none	MnO_2	PbO_2	NaCl
Volume O_2 (cm³)	10	180	160	12

(a) Which of the solids are catalysts?   **(1 mark)**

MnO_2 and PbO_2

Only these two solids **significantly** increase the rate of reaction.

(b) How would you show that one of the solids identified in (a) was a catalyst and not a reactant?   **(4 marks)**

Weigh the solid at the start of the reaction. After the reaction, filter and dry the solid remaining. Re-weigh the solid, which should have the same mass as at the start.

The solid at the end of the reaction must be shown not to have been used up.

Now try this

Describe briefly an experiment to show if any of three solids, **A**, **B** and **C**, are a catalyst for the decomposition of hydrogen peroxide, H_2O_2. The equation for the reaction is:

$$2H_2O_2 \rightarrow 2H_2O + O_2$$

(4 marks)

Practical skills   Remember that this must be a fair test. Describe how you would monitor the rate of reaction, and how you would see whether each solid increased the rate of reaction.

Reversible reactions

Some reactions are **reversible**. This means that the products can react to make the original reactants.

Equations

Many reactions are not reversible. They go to completion and the products cannot react with each other.

$$A + B \rightarrow C + D$$

the reactants   the products

Reversible reactions are shown with a split arrow pointing in both directions:

$$A + B \rightleftharpoons C + D$$

shows that the reaction is reversible

A reversible reaction

Ammonium chloride is a white solid. It breaks down when heated, forming ammonia and hydrogen chloride gases:

$$NH_4Cl \rightarrow NH_3 + HCl$$

When the products cool down, they react together to form ammonium chloride:

$$NH_3 + HCl \rightarrow NH_4Cl$$

Overall, the reaction is shown like this:

$$NH_4Cl \underset{cool}{\overset{heat}{\rightleftharpoons}} NH_3 + HCl$$

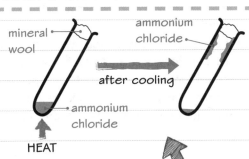

mineral wool   ammonium chloride

after cooling

ammonium chloride

HEAT

The mineral wool stops ammonium chloride escaping; the solid forms where the test tube is cool.

Energy changes in reversible reactions

- A reversible reaction that is exothermic in one direction will be endothermic in the opposite direction.
- amount of energy released in one direction = amount of energy absorbed in opposite direction
- example: hydrated copper sulfate $\underset{exothermic}{\overset{endothermic}{\rightleftharpoons}}$ anhydrous copper sulfate + water
  BLUE   WHITE

Worked example

Ammonia is made from nitrogen and hydrogen in a reversible reaction: $N_2 + 3H_2 \rightleftharpoons 2NH_3$
The table shows the yield of ammonia at different temperatures.

Temperature (°C)	% yield of ammonia
450	38
500	30
550	16

(a) What happens to the yield of ammonia as the temperature is increased? **(1 mark)**

It decreases.

(b) Give **two** reasons why ammonia is **not** manufactured at a very high temperature. **(2 marks)**

The yield is greater at low temperatures and less energy will be needed for the process.

Now try this

When hydrated copper sulfate is heated it turns white. Explain why this occurs and how this reaction can be used as a test for the presence of water. **(4 marks)**

59

Equilibrium

Reversible reactions (described on page 59) can reach **equilibrium** if they occur in apparatus where the reactants and products cannot escape.

Equilibrium example

- Hydrogen and iodine gases are mixed in a container.
- The container is sealed to make a **closed system**.
- The **forward reaction** $H_2 + I_2 \rightarrow 2HI$ and **backward reaction** $2HI \rightarrow H_2 + I_2$ continue all the time.
- Equilibrium is reached when the rates of the forward and the backward reactions are equal. At this point, the concentrations of all the reactants and products do not change any further.

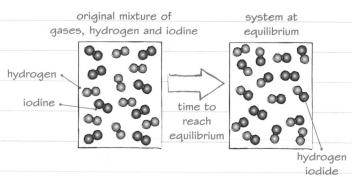

original mixture of gases, hydrogen and iodine

system at equilibrium

hydrogen

iodine

time to reach equilibrium

hydrogen iodide

$$H_2(g) + I_2(g) \rightleftharpoons 2HI(g)$$

Three positions of equilibrium

- At equilibrium the **rates** of the forward and backward reactions are equal.
- The **amounts** of the reactants and products do **not** have to be equal.
- In the example, the boxes show three possible positions of equilibrium.

$$H_2(g) + I_2(g) \rightleftharpoons 2HI(g)$$

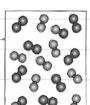 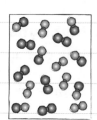 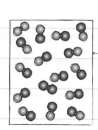

position of equilibrium is far to the left (mainly reactants)

position of equilibrium is far to the right (mainly products)

Worked example

Hydrogen and hydrogen iodide are colourless gases. Iodine is a purple gas. Explain what you would **see** if the position of equilibrium between hydrogen, iodine and hydrogen iodide moves to the left. **(2 marks)**

If the position of equilibrium moves to the left, more iodine is formed. The mixture would look darker purple.

Now try this

If the reaction between hydrogen, iodine and hydrogen iodide is at **equilibrium**:

A the amount of each gas is equal

B the reactions have stopped

C the rate of the forward and the backward reactions are equal

D the volume of each gas is equal. **(1 mark)**

Extended response – Rates of reaction

You can revise the information for this question, which is about **rates of reaction**, on pages 52–58.

Worked example

Marble chips react with dilute hydrochloric acid.

$$CaCO_3 + 2HCl \rightarrow CaCl_2 + H_2O + CO_2$$

Explain how you could use the apparatus shown below to investigate how the size of the marble chips affects the rate of this reaction. **(6 marks)**

Basic method

- Measure out some dilute hydrochloric acid using a measuring cylinder.
- Weigh out some marble chips.
- Add the chips to the acid and attach the gas syringe.
- Measure the volume of carbon dioxide gas given off each minute for ten minutes.

Changing surface area

- Repeat the experiment with the same volume of acid and same mass of marble chips.
- Use large, medium, small and very small chips.

Analysing results

- See how much gas is given off by the different sized chips.
- The smaller the marble chips, the more gas will be given off in ten minutes.

The answer has been well structured into three sections. The names of the pieces of apparatus have been used.

The acid and the marble chips have been measured out so that a fair comparison can be made.

Although the **size** of the marble chips is altered, the **total mass** of the chips is the same each time.

There are different ways of analysing the results. Another way is to draw a graph.

Now try this

In a similar reaction to that described above, with large chips, medium chips and small chips, the results are plotted on the graph. Explain the graph shown. **(6 marks)**

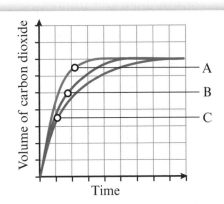

Crude oil

When sea life dies and is buried under mud over a very long period of time, it may turn into the liquid **crude oil**.

Fractional distillation

Crude oil is a mixture of a very large number of compounds. These can be separated in industry into groups of molecules with similar numbers of carbon atoms, called **fractions**, by **fractional distillation**. (You can revise laboratory fractional distillation on page 3.)

Fractional distillation of crude oil happens in a fractionating column:
- The oil is heated to evaporate it.
- Vapour from the oil rises up the column.
- Each fraction condenses at a different temperature because it has a different range of boiling points.

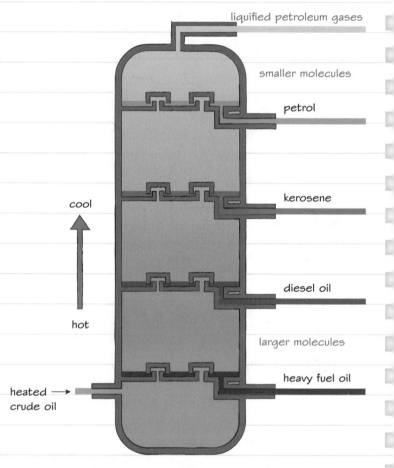

Uses of products from crude oil

Many of the useful materials that we use in everyday life are produced using the fractions from crude oil.

1 The fractions can be processed to make different **fuels**, shown in the diagram.

2 Some fractions are a **feedstock** (starting material) for the **petrochemical industry** to make other products, such as **solvents**, **lubricants**, **polymers** and **detergents**.

You can revise polymers on page 70.

Worked example

(a) Describe how crude oil is extracted from the Earth. **(2 marks)**

Crude oil is found in rocks. It is extracted by drilling into the rocks.

(b) Explain why it is important to conserve supplies of crude oil. **(2 marks)**

Crude oil is **finite**. Once humans use it up, it cannot be replaced.

Now try this

The diagram shows the formula and boiling point of some molecules in three fractions.

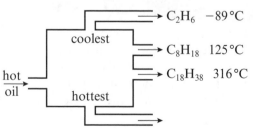

How does the number of carbon atoms in a molecule affect its boiling point? **(1 mark)**

Alkanes

Crude oil is a mixture of a very large number of compounds, most of which are hydrocarbons.

Alkanes

A **hydrocarbon** is a compound made up of hydrogen atoms and carbon atoms only. Most of the hydrocarbon molecules in crude oil are **alkanes**.

Alkanes have the general formula C_nH_{2n+2}. For example, the chemical formula for butane (which contains four carbon atoms) is C_4H_{10}.

A **compound** consists of two or more elements chemically combined together. Take care **not** to write that hydrocarbons are mixtures of hydrogen and carbon.

Take care when writing the formulae for alkanes: C^4H^{10} or C4H10 would be wrong.

Formulae of first three alkanes

An alkane molecule can be represented by its formula or its displayed structure, showing all the bonds.

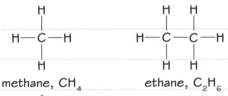

methane, CH_4 ethane, C_2H_6 propane, C_3H_8

The atoms in hydrocarbon molecules are joined together by **covalent bonds**.

The names of alkanes end in 'ane'.

Each hydrogen atom has one bond and each carbon atom has four bonds.

Worked example

Butane is an alkane with four carbon atoms. Draw its displayed structure. **(2 marks)**

When a structure is drawn out like this one, each line represents a covalent bond. Carbon atoms each form four covalent bonds, and hydrogen atoms each form only one bond. Remember that covalent bonds form when atoms share electrons.

You need to know the names of the four alkanes on this page.

Alkanes are saturated hydrocarbons – their carbon atoms are joined together by single covalent bonds only.

Now try this

1 Crude oil contains saturated hydrocarbons. State what is meant by:
   (a) saturated   **(1 mark)**
   (b) hydrocarbon.   **(1 mark)**
2 State the general formula for alkanes. Use n for the number of carbon atoms. **(1 mark)**
3 Compound X has the formula C_5H_{12}. State why X is an alkane. **(1 mark)**
4 Hexane is an alkane with six carbon atoms. Draw its displayed structure. **(2 marks)**

Properties of hydrocarbons

Some of the properties of the alkane molecules found in crude oil depend on the size of the molecules.

Worked example

The boiling points of the first four alkanes are given.

Alkane	Boiling point (°C)
methane	−162
ethane	−89
propane	−42
butane	0

(a) Plot these points on a graph.   **(3 marks)**

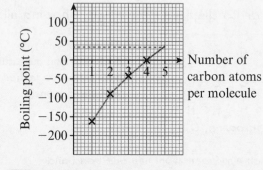

Maths skills Choose an even scale, and plot the points carefully, especially with the negative values.

(b) What pattern is shown in the data?   **(1 mark)**

The boiling point increases with molecular size.

(c) Use your graph to estimate the boiling point of a five-carbon alkane.   **(1 mark)**

36 °C

Boiling point, viscosity and flammability

There are trends in the properties of the different fractions from crude oil.

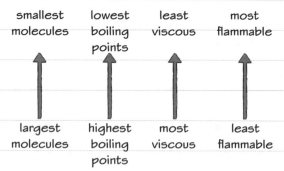

These properties influence how the fractions are used as fuels.

The less viscous and more flammable fractions (e.g. petrol) can be used in small vehicles such as cars. The more viscous and less flammable fractions (e.g. heavy fuel oil) can be used in large-scale boilers, e.g. in factories.

Maths skills Extend your graph beyond the plotted points to find where the graph reaches. This is called **extrapolation**.

Hydrocarbon fuels

With plenty of air, hydrocarbon fuels undergo complete combustion:
- Carbon atoms are oxidised to CO_2.
- Hydrogen atoms are oxidised to H_2O.
- Lots of heat is released.

Maths skills Combustion example

$$C_3H_8 + 5O_2 \rightarrow 3CO_2 + 4H_2O$$

In this equation, oxygen atoms are in three different substances, so leave the balancing of these until last.

1 Balance the C: 3 on left, so $3CO_2$ on right.

2 Balance the H: 8 on left, so $4H_2O$ on right.

3 Finally, balance the O: 10 on right (6 in $3CO_2$ and 4 in $4H_2O$), so $5O_2$ on left.

Now try this

1 Explain how the molecules found in the liquefied petroleum gas fraction are different to those in the heavy fuel oil fraction, and why they are extracted from a different place in the fractional distillation tower.   **(3 marks)**

2 State the property of crude oil fractions that allows them to be separated during fractional distillation.   **(1 mark)**

3 Describe how crude oil is separated by fractional distillation.   **(2 mark)**

Cracking

There is a shortage of the hydrocarbon fractions with small molecules that are in high demand for fuels. Some hydrocarbons with larger molecules are broken down (**cracked**) to produce these smaller, more useful molecules.

Cracking

Cracking is a **thermal decomposition** reaction. In cracking, oil fractions are heated so they vaporise. Their vapours are either:
• passed over a hot catalyst, or
• mixed with steam and heated to very high temperatures.

Hydrocarbons in modern life

• alkanes used as fuels
• alkanes used as feedstock to make other products, such as solvents, lubricants and detergents
• alkenes used to make polymers
• alkenes used as a feedstock to make many other chemicals

Products of cracking

Cracking a larger alkane molecule produces:

1 a shorter alkane molecule

2 a molecule containing a C=C double bond (an **alkene**) (You can revise alkenes on page 66.)

Worked example

(a) Complete the balanced equation for the cracking of the hydrocarbon C_6H_{14} to produce ethene, C_2H_4, and an alkane. **(2 marks)**

$$C_6H_{14} \rightarrow C_2H_4 + C_4H_{10}$$

(b) Suggest **two** reasons why there is a greater demand for the products than for the original hydrocarbon. **(2 marks)**

Smaller hydrocarbons make better fuels than larger ones. Alkenes are used to make polymers.

Remember:
• alkanes – the number of H atoms is double the number of C atoms plus two.

This reaction is also possible:
$C_6H_{14} \rightarrow 2C_2H_4 + C_2H_6$

Compared with larger hydrocarbons, smaller hydrocarbons are less viscous (more runny), more flammable and have lower boiling points (they tend to be gas or liquid). They are more useful as fuels.

Now try this

1 The diagram shows an experiment to crack paraffin soaked onto mineral wool.
   (a) Explain why the paraffin is warmed. **(1 mark)**
   (b) What type of reaction occurs in the cracking process? **(1 mark)**
   (c) C_2H_4 molecules are produced. State whether they would collect at position X or position Y. **(1 mark)**

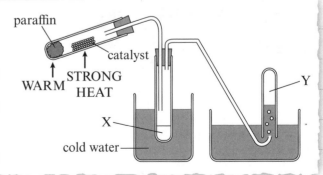

Alkenes

There are a very large number of natural and **synthetic** carbon compounds. This is because carbon atoms can form stable chains of various lengths. One **homologous series**, or family of compounds, is the alkanes (described on page 63). Other homologous series are the **alkenes** (this page), the alcohols (page 68) and the carboxylic acids (page 69).

Alkenes

- Alkenes are produced by cracking long-chain alkanes. → You can revise cracking on page 65.
- Alkenes are hydrocarbons with a carbon-to-carbon double bond.
- The general formula for alkenes is C_2H_{2n}.
- Alkenes are **unsaturated**.

As alkenes are unsaturated, an alkene molecule with one C=C bond will have two fewer hydrogen atoms than the alkane molecule with the same number of carbon atoms. Alkane molecules are **saturated**.

Formulae of first three alkenes

ethene, C_2H_4 propene, C_3H_6 butene, C_4H_8

Worked example

Pentene is an alkene with five carbon atoms. Draw its displayed structure.   **(2 marks)**

The C=C double bond has been shown between the first two carbons in the molecule, but another version of pentene has the C=C in the centre of the molecule.

You need to know the names of the four alkenes on this page.

Pentene is an unsaturated molecule. It has 10 hydrogen atoms, but the alkane with 5 carbon atoms has 12 hydrogen atoms.

Now try this

Consider molecules **A**, **B**, **C** and **D**.

A   **B**   **C**   **D**

Which of the molecules is (or are):

(a) a hydrocarbon   **(1 mark)**   (b) unsaturated   **(1 mark)**   (c) an alkane   **(1 mark)**

Reactions of alkenes

Members of a homologous series have the same **functional group**. The reactions of this functional group will be the same for each member of homologous series. For alkenes, the functional group is the C=C bond. This double bond makes alkenes more reactive than alkanes.

Addition reactions with hydrogen, halogens and water

In an alkene, the carbon–carbon double bond can become a carbon–carbon single bond, and an atom can then bond to each of the carbons that were in the double bond.

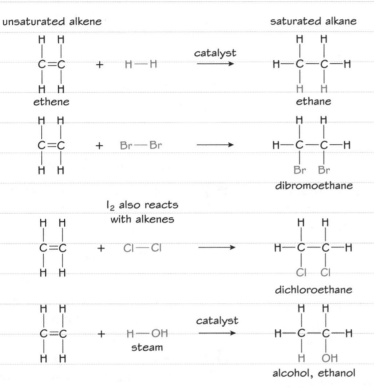

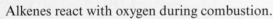

Worked example

Alkenes react with oxygen during combustion.

(a) Write the equation, using displayed formulae, for the complete combustion of butene. **(3 marks)**

$C_4H_8 + 6O_2 \rightarrow 4CO_2 + 4H_2O$

(b) Explain why a burning alkene may give a smoky flame. **(2 marks)**

The smoky flame is caused by soot (unburned carbon) and incomplete combustion.

> Complete means full oxidation to make CO_2 and H_2O.

> The C=C in alkenes is harder to break than the C–C bonds in alkanes, so alkenes are more likely to burn with a smoky flame.

Now try this

Iodine reacts with alkenes in a similar way to bromine and chlorine.

(a) Write the equation, using displayed formulae, for the reaction between propene and iodine. **(3 marks)**

(b) Name the product formed in (a). **(1 mark)**

(c) Explain how you could distinguish between propane and propene. **(3 marks)**

> **Practical skills** Use the reaction with bromine to distinguish between propane and propene. The bromine solution used is orange. When it reacts with an alkene, the product is colourless.

Alcohols

Alcohols are used as solvents and fuels. **Ethanol** is the main alcohol in alcoholic drinks such as wine and beer.

Structures

Alcohols all contain the **functional group** –OH.

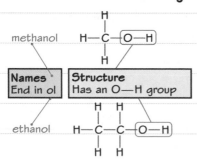

methanol

Names End in ol | **Structure** Has an O—H group

ethanol

The alcohols form a **homologous series**. They have:
- the same functional group
- similar chemical properties.

You should be able to recognise alcohols from their names or formulae:
- methanol is CH_3OH
- ethanol is CH_3CH_2OH
- propanol is $CH_3CH_2CH_2OH$
- butanol is $CH_3CH_2CH_2CH_2OH$

Reactions of alcohols

Methanol, ethanol, propanol and butanol all:
- dissolve in water to form a neutral solution
- react with sodium to produce hydrogen
- burn in air to produce carbon dioxide and water.

For example:

methanol + oxygen → carbon dioxide + water

$$2CH_3OH(l) + 3O_2(g) \rightarrow 2CO_2(g) + 4H_2O(l)$$

- can be oxidised to produce carboxylic acids (see page 69)
- react with carboxylic acids to make esters (see page 69).

Formation of ethanol

1 A solution of ethanol can be produced by **fermentation** of sugar solutions:

✓ Start with a sugar solution.
✓ Add **yeast**.
✓ Keep in a warm place.
✓ Prevent any oxygen getting into the apparatus (**anaerobic** conditions).

2 Ethanol can also be produced by the addition reaction of steam with ethene, using a **catalyst** (revise addition reactions of alkenes on page 67).

Worked example

The diagram shows the structure of propanol.
Explain how you know from its structure that propanol is an alcohol and not an alkane. **(2 marks)**

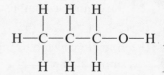

Alkanes are hydrocarbons, so they only contain hydrogen and carbon atoms, but propanol contains an oxygen atom in its –OH group. The –OH group shows that this molecule is an alcohol.

Now try this

1 You are given a solution of ethanol and a solution of ethanoic acid. How can you tell which is which? **(2 marks)**

2 (a) Draw the structure of a molecule of butanol, showing all covalent bonds. **(1 mark)**

(b) Suggest a use for the liquid alcohol, hexanol. **(1 mark)**

3 Balance this equation for the combustion of propanol:

$$2C_3H_7OH + 9O_2 \rightarrow \ldots CO_2 + \ldots. \quad \textbf{(2 marks)}$$

Carboxylic acids

Carboxylic acids can be produced by the oxidation of alcohols.

Structures

Carboxylic acids contain the **functional group** –COOH.

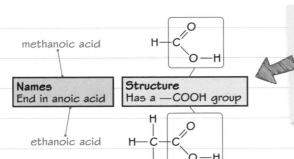

methanoic acid

| Names | Structure |
| End in anoic acid | Has a —COOH group |

ethanoic acid

You should be able to recognise carboxylic acids from their names or formulae:
• methanoic acid is $HCOOH$
• ethanoic acid is CH_3COOH
• propanoic acid is CH_3CH_2COOH
• butanoic acid is $CH_3CH_2CH_2COOH$.

Reactions of carboxylic acids

Carboxylic acids:
• dissolve in water to form acidic solutions
• react with carbonates to produce carbon dioxide
• react with alcohols to produce **esters** when an acid catalyst is added. For example, ethanoic acid + ethanol → ethyl ethanoate + water

 Acids make solutions acidic because they release hydrogen ions, $H^+(aq)$.

Worked example

The diagram shows the structure of propanoic acid.
Describe **two** tests you could do to show that propanoic acid solution is acidic. **(4 marks)**

1 Add universal indicator. The mixture should go red or orange, showing that it is acidic.

2 Add it to some calcium carbonate and look for fizzing. The gas given off should turn limewater cloudy.

Now try this

 1 What is the empirical formula of butanoic acid? **(1 mark)**
    A $CH_3CH_2CH_2COOH$
    B $C_4H_8O_2$
    C C_3H_7COOH
    D C_2H_4O

 2 (a) Explain how you know that butanoic acid, C_3H_7COOH, is a carboxylic acid. **(2 marks)**
    (b) Name the type of substance formed when butanoic acid reacts with ethanol. **(1 mark)**

The empirical formula is the simplest ratio of atoms of carbon, hydrogen and oxygen in the formula of butanoic acid.

1. Count up all the atoms of each element.

2. Can these numbers be simplified?

Polymers

Polymers are long-chain molecules that occur naturally (see page 71) or can be manufactured. Synthetic polymers made in addition reactions are **addition polymers**.

Addition polymerisation

Alkene molecules can act as **monomers**. They can join together in a **polymerisation** reaction to form very large molecules (long chains) called **polymers**.

For example, ethene forms poly(ethene).

monomers
polymer

$$n \; C{=}C \rightarrow {+}C{-}C{+}_n$$

The displayed structures show that many (*n*) ethene monomers can react together to form one molecule of poly(ethene).

Worked example

(a) Draw the displayed structure of the polymer that forms from propene. **(1 mark)**

$$n \; C{=}C \rightarrow {+}C{-}C{+}_n$$
(H, H / H, CH₃ → H, H / H, CH₃)

(b) Name the polymer formed. **(1 mark)**

poly(propene)

The name of a polymer is given by its monomer – it is poly(name of monomer).

You should be able to show the formation of a polymer from a given alkene monomer. Students have often struggled to do this in exams. They often forget to change the double bond to a single bond or they leave out the *n*.

To convert from a monomer to a polymer:
• draw the monomer but with a single bond
• draw a long bond either side
• draw brackets through the long bonds
• write *n* after the bracket.

Worked example

Draw the monomer that forms the polymer with this formula:

(1 mark)

$${+}C{-}C{+}_n \qquad C{=}C$$
(F, F / F, F)   (F, F / F, F)

To go from the formula to the monomer:

1 remove the long bonds, brackets and *n*

2 join the two carbons that the long bonds came from with a double bond.

Characteristics of addition polymerisation

1 The monomer has a C=C double bond.

2 The repeating unit has the same number and type of atoms as the monomer.

3 Only the polymer forms with no other product.

Now try this

1 (a) Complete this equation to show how a polymer forms from chloroethene. **(1 mark)**

$$n \; C{=}C \rightarrow$$
(H, H / H, Cl)

(b) Name the polymer formed. **(1 mark)**

2 Explain whether poly(ethene) is a saturated or an unsaturated hydrocarbon. **(2 marks)**

DNA

Some polymers occur naturally in the human body.

Naturally occurring polymers

These four naturally occurring polymers are found in living things.

1 proteins (monomers: amino acids)

2 DNA (monomers: nucleotides)

3 starch (monomers: sugars)

4 cellulose (monomers: sugars)

 Proteins are found in cells and carry out many of the functions of a cell.

 Starch is a food store in plants.

 Cellulose is used in the cell walls of plant cells.

Deoxyribonucleic acid

A deoxyribonucleic acid, DNA, molecule is a **double helix**.

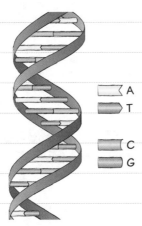

A
T
C
G

- DNA is a large molecule essential for life.
- DNA encodes genetic instructions for the development of living organisms and viruses.
- The monomers that make up DNA are called **nucleotides**.

Worked example

Which statement is true about polymers found in living organisms?
(1 mark)

A A large number of identical amino acids join to form a protein.
B DNA is a polymer chain made from four amino acids.
C Sugar monomers can polymerise to form the polymers starch and cellulose.
D Four nucleotides join to make a DNA monomer.

C

There are many technical terms to know on this page! **A** is incorrect because proteins are formed from different monomers. **B** is incorrect because DNA monomers are nucleotides. **D** is incorrect because DNA is a polymer.

Now try this

What monomers make up the polymer DNA? **(1 mark)**
A alkenes   **C** nucleotides
B amino acids   **D** sugars

Extended response – Organic chemistry

You can revise the information for this question, which is about **cracking**, on pages 62 and 65.

Worked example

Fractional distillation of crude oil produces different amounts of the fractions. The graph shows the supply of, and demand for, three of these fractions.

Explain the problem caused by the difference in the supply and demand of fractions, and explain how an oil refinery can solve this problem. **(6 marks)**

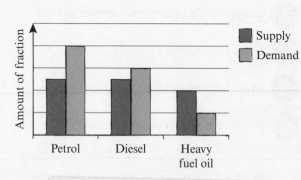

There is a shortage of the fractions with smaller molecules (petrol and diesel). There is a surplus of the fraction with larger molecules (heavy fuel oil).

This means that there would not be enough of the fractions with smaller molecules. There would be too much of the fraction with larger molecules (it might be wasted, remain unsold or be sold at a very low price).

The larger molecules can be changed into shorter molecules by cracking. The molecules are heated, and the hot vapour is passed over a catalyst. Alternatively, the molecules are mixed with steam and heated to a very high temperature.

The molecules decompose, making shorter alkane molecules that are in higher demand as fuels.

Cracking can also make alkenes which can be used to make polymers.

 First, use the chart to identify the problem of supply and demand.

 There are two methods of cracking – using a catalyst or mixing with steam and using a higher temperature.

 Cracking produces both shorter alkane molecules and alkene molecules. Give the main use for each type of molecule to explain why cracking is so useful.

 You could give example equations in this answer, for example:

$$C_{20}H_{42} \rightarrow C_{12}H_{24} + C_8H_{18}$$

This equation is easy to construct, because alkanes have the general formula C_nH_{2n+2} and alkenes have the general formula C_nH_{2n}. Write on the left the formula of any alkane and on the right any alkene with fewer carbons. You can balance the equation with an alkane, which you can work out by balancing the carbon atoms and hydrogen atoms on each side.

Now try this

You are provided with three unlabelled tubes containing colourless liquids. One contains an alkene, one an alcohol and one the solution of a carboxylic acid. Describe test tube reactions that you could carry out to identify each substance. **(6 marks)**

 **Practical skills** There are different reactions of alkenes, alcohols and carboxylic acids. You need to choose just one for each compound that you can easily carry out in a laboratory.

72

Pure substances and formulations

A pure substance is a single element or a single compound. Some mixtures are designed as useful products, and are called **formulations**.

 **Practical skills** · **Testing purity**

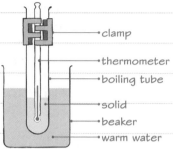

Measuring melting point

The liquid in the beaker is heated gently, and the temperature is taken when the solid melts.

👍 A pure element or pure compound melts and boils at a specific temperature.

👍 The measured melting point or boiling point will match that in a data book.

👎 Mixtures have variable melting points and boiling points which are not sharp.

Elements, mixtures and compounds can be revised on page 1.

Formulations

A formulation is a mixture, often complex, designed for a specific use:

- Each component in the mixture has a specific purpose.
- The components are added in measured amounts.
- Examples are fuels, cleaning agents, paints, medicines, alloys, fertilisers and foods.

Separating mixtures

Mixtures can be separated using the physical properties of the components.

- You can revise filtration, crystallisation and chromatography on page 2.
- You can revise distillation on page 3.
- Further experimental details about chromatography are on the next page.

Worked example

Three students were preparing aspirin. In the last stage, the aspirin was filtered then dried. The students then found the melting point.

Melting point results (°C)

Student A	Student B	Student C	Data book
124	132–136	135	136

Evaluate the data. **(6 marks)**

- Student A's sample has a sharp melting point, but lower than the data book value. It is a pure substance, but it is not aspirin.

- Student B's sample does not have a sharp melting point, but goes up to the data book value. It may be impure aspirin – the impurity may be water if the aspirin was not completely dried.

- Student C's sample has a sharp melting point that is only 1 °C from the data book value so it is probably pure (dry) aspirin.

Purity

Scientists use the word pure to mean a single element or compound. In everyday life, 'pure' can mean a substance that has nothing added (it is **unadulterated**).

Example: milk
- In everyday life, pure milk means milk with nothing added to it.
- In scientific terms, milk is a mixture, including water and fat.

Now try this

The diagram shows the components in a typical watercolour paint.

(a) Explain why this paint can be described as a formulation. **(3 marks)**

(b) Suggest reasons for any **two** of the components being in the paint. **(2 marks)**

pigment
brightener
glue
additives
thickener
water

Core practical – Chromatography

Core practical

Aim

To identify a substance by carrying out paper chromatography and measuring the R_f value.

Apparatus

- eye protection
- beaker
- chromatography paper
- pencil and ruler
- solvent
- coloured substances
- pipettes

Method

1. Use a ruler to draw a pencil line on the chromatography paper.
2. Use a pipette to put a drop of the original coloured mixture on the pencil line.
3. Place some solvent in a beaker, and place the chromatography paper in the solvent.
4. Allow the solvent to rise up the paper, so the substances in the original mixture separate.

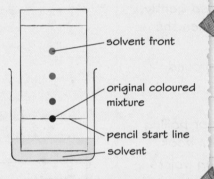

solvent front

original coloured mixture

pencil start line

solvent

Sample results

Distance travelled in mm

Solvent front	Red substance	Green substance	Blue substance
75	60	27	12

$R_f = \dfrac{\text{distance moved by substance}}{\text{distance moved by solvent}}$

R_f (red) = 60/75 = 0.80
R_f (green) = 27/75 = 0.36
R_f (blue) = 12/75 = 0.16

In this experiment, the chromatography paper is the **stationary phase**. The solvent, which moves up the paper and carries with it the components of the mixture, is the **mobile phase**.

The different substances in a mixture are separated because they have different attractions to the stationary phase and the mobile phase. In the experiment shown, the red substance has the highest attraction to the solvent and/or lack of attraction to the paper.

Some substances may have the same solubility in a solvent (and so make one spot). A pure substance will produce only a single spot in **all** solvents. Different coloured substances have different R_f values.

The R_f values depend on the solvent used.

The R_f values for different substances in the solvent used could be looked up in a data book, and used to identify the substances.

Now try this

A chromatography experiment is set up.
- An ink pen is used to draw a start line on the chromatography paper.
- A pipette is used to put a drop of the original coloured mixture on the start line.
- Some solvent is placed in a beaker, and the chromatography paper is placed so that the start line is under the solvent.
- The solvent is allowed to rise until it is at the top of the paper.

R_f values are calculated using the formula:

$R_f = \dfrac{\text{distance moved by substance}}{\text{length of paper}}$

- Identify **three** errors in this method and suggest improvements. **(6 marks)**

Tests for gases

🧪 **Practical skills** The gases hydrogen, oxygen, carbon dioxide and chlorine can be identified using simple tests.

Test for hydrogen

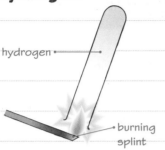

- Light a splint.
- Place burning splint by the gas.
- 👍 Hydrogen burns rapidly with a squeaky pop sound.

Test for oxygen

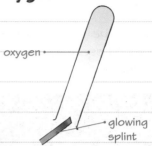

- Light a splint.
- Blow out flame but leave glowing end to splint.
- Place glowing splint into the gas.
- 👍 Oxygen causes splint to relight.

Test for carbon dioxide

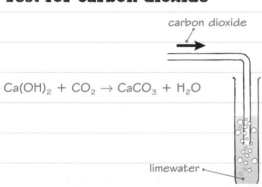

$$Ca(OH)_2 + CO_2 \rightarrow CaCO_3 + H_2O$$

- Place some calcium hydroxide solution (limewater) into a test tube.
- Bubble the gas into the limewater.
- 👍 Carbon dioxide causes the limewater to go milky (cloudy).

Test for chlorine

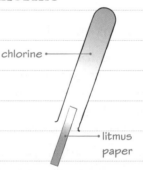

- Dampen some litmus paper.
- Place the litmus paper into the gas.
- 👍 Chlorine causes red litmus paper to bleach (go white).
- 👍 Chlorine causes blue litmus paper to go red then bleach (go white).

Worked example

A compound of sodium is added to dilute hydrochloric acid. It gives off a gas that is bubbled through limewater. The limewater goes milky.

(a) Identify the gas given off. **(1 mark)**

carbon dioxide

(b) Write the word equation for the reaction of this gas with limewater and identify the milky substance. **(3 marks)**

calcium hydroxide + carbon dioxide → calcium carbonate + water
The milky substance is calcium carbonate.

(c) Identify the original compound. **(1 mark)**

sodium carbonate

Now try this

A lighted splint is put into a boiling tube of gas. The gas goes pop and some condensation is seen at the top of the boiling tube.
Explain these observations. **(3 marks)**

Once you have identified the gas, think what reaction happens when the gas is ignited.

What type of compound reacts with acids to give off carbon dioxide gas?

Tests for cations

 Practical skills Some positive ions (cations) can be identified by a flame test or by their reaction with sodium hydroxide solution.

Flame tests

Flame tests can be used to identify some metal ions (cations). These ions produce distinctive colours in a flame.

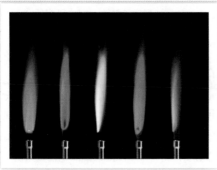

calcium   sodium   potassium
lithium   copper

 Practical skills

1 Dip a flame test wire loop in acid then heat in a blue Bunsen flame until there is no colour.

2 Dip the loop in acid again and then into the compound.

3 Hold the loop in the Bunsen flame.

An alternative method is to dip a damp splint into the solid and place this in the flame. This avoids the use of a corrosive acid.

Hydroxide precipitates

Some metal ions form coloured hydroxide **precipitates**. The sample solution is placed in a test tube and a few drops of dilute sodium hydroxide solution are added. The table shows the colours you need to know.

Some metal ions form white precipitates with sodium hydroxide solution:
- magnesium ions, Mg^{2+}
- calcium ions, Ca^{2+}
- aluminium ions, Al^{3+}.

Metal ion	Colour of precipitate
copper(II), Cu^{2+}	blue
iron(II), Fe^{2+}	green
iron(III), Fe^{3+}	brown

Copper and iron are **transition metals**. Many of these elements form coloured compounds. Although sodium hydroxide is soluble in water, most hydroxides are insoluble and so form precipitates in these tests.

Worked example

Describe how to distinguish between the hydroxide precipitate formed by aluminium ions and the one formed by magnesium ions. **(2 marks)**

The aluminium hydroxide precipitate will dissolve in excess sodium hydroxide but the magnesium hydroxide precipitate will not.

Calcium ions also form a hydroxide precipitate that does not dissolve in excess sodium hydroxide solution.

Now try this

Calcium in the diet is important for the development of healthy bones and teeth.
(a) Describe **two** ways in which calcium ions may be detected. **(4 marks)**
(b) Explain why it is not possible to distinguish magnesium ions from calcium ions in one of these tests. **(2 marks)**

Tests for anions

 Practical skills   Non-metal ions (anions) can be identified using simple laboratory tests. When combined with tests for metal ions, unknown compounds can be identified.

Tests for carbonate, sulfate and halide ions

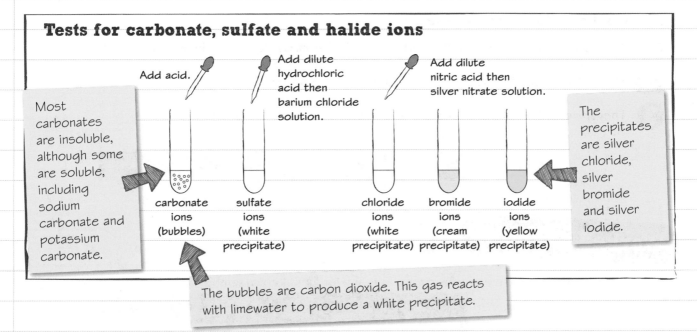

Add acid.

Add dilute hydrochloric acid then barium chloride solution.

Add dilute nitric acid then silver nitrate solution.

Most carbonates are insoluble, although some are soluble, including sodium carbonate and potassium carbonate.

carbonate ions (bubbles)

sulfate ions (white precipitate)

chloride ions (white precipitate)

bromide ions (cream precipitate)

iodide ions (yellow precipitate)

The precipitates are silver chloride, silver bromide and silver iodide.

The bubbles are carbon dioxide. This gas reacts with limewater to produce a white precipitate.

Worked example

A white powder is used to treat water for drinking, and wastewater before it is discharged into rivers.

It produces a white precipitate with sodium hydroxide solution, which dissolves in excess alkali. It does not bubble when dilute hydrochloric acid is added to it, but produces a white precipitate when barium chloride is then added. Explain what the name and formula of the powder is. **(4 marks)**

The test with sodium hydroxide solution shows that the powder contains aluminium ions. The test with barium chloride shows that it contains sulfate ions. It must have been aluminium sulfate, $Al_2(SO_4)_3$.

Now try this

 1 A cargo of clothing was damaged by water while at sea. It could have been caused by rainwater or by seawater. The table shows the results of tests using silver nitrate solution. Explain what the tests show. **(3 marks)**

Sample	Test result
rainwater	no visible change
seawater	white precipitate
damp from clothing	white precipitate

 2 The labels have come off four bottles. The bottles contain aluminium sulfate, magnesium sulfate, potassium iodide and potassium bromide. Describe how you would use chemical tests to identify the contents of each bottle. **(6 marks)**

You will need to refer to the tests on this page and the previous page.

Flame emission spectroscopy

Metal ions can be detected using simple chemical tests (flame tests and precipitation reactions), described on page 76, or by flame emission spectroscopy.

Flame emission spectroscopy

1 The sample, a solution containing one or more metal ions, is placed in a flame.

2 The light emitted is passed through a **spectroscope**.

3 The spectroscope produces a line spectrum, unique for each ion.

sodium ions, Na⁺

Worked example

The diagram shows the line spectrum for two ions and for an unknown sample.

calcium ions, Ca²⁺

magnesium ions, Mg²⁺

unknown sample

Use the diagram and the line spectrum for sodium ions, above, to describe the composition of the unknown sample. **(2 marks)**

The unknown sample has emission lines that match sodium ions and calcium ions. It is a mixture containing these two ions.

Instrumental methods

Instrumental methods use a machine to automatically carry out a measurement, rather than a scientist carrying out a laboratory test.

Flame emission spectroscopy is an instrumental method and, compared with anion and cation tests (pages 76 and 77), is:

👍 more accurate

👍 more sensitive (only a tiny sample is needed)

👍 fast.

Flame emission spectroscopy is ideal when only a small sample is available, which would be too small to give a visible flame colour or precipitate.

Now try this

Give the advantages of using flame emission spectroscopy, apart from speed and sensitivity, to distinguish between the following samples:

(a) calcium chloride and magnesium chloride, compared with using their reaction with sodium hydroxide solution **(2 marks)**

(b) lithium chloride and calcium chloride, compared with using flame tests. **(4 marks)**

What would you see if lithium chloride and calcium chloride were each tested in a flame test? Could there be a problem if the samples were impure?

Core practical – Identifying a compound

Core practical

Aim

To identify an ionic compound, using laboratory tests for cations and anions.

> The experimenter does not know the identity of the compound, so it should be assumed to be corrosive and toxic, unless told otherwise.

Apparatus

- eye protection and gloves
- sample of unknown solid ionic compound
- flame test wire loops
- test tubes and rack
- Bunsen burner
- distilled water
- pipettes
- spatula
- delivery tube
- sodium hydroxide solution
- limewater
- dilute hydrochloric acid
- dilute nitric acid
- silver nitrate solution
- barium chloride solution

Method

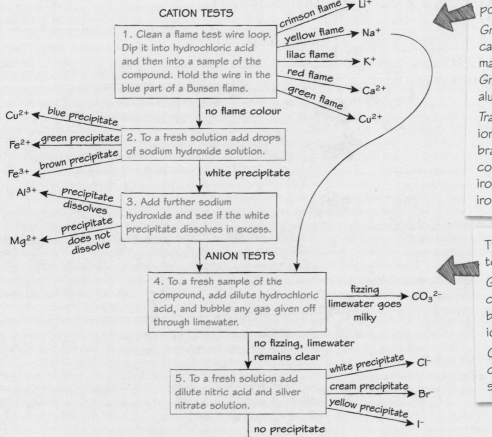

> The cations that you need to know are:
>
> Group 1 (1+ ions)
> lithium, Li^+
> sodium, Na^+
> potassium, K^+
>
> Group 2 (2+ ions)
> calcium, Ca^{2+}
> magnesium, Mg^{2+}
> Group 3 (3+ ion)
> aluminium, Al^{3+}
>
> Transition metals (variable ions, the charge is given in brackets)
> copper(II), Cu^{2+}
> iron(II), Fe^{2+}
> iron(III), Fe^{3+}

> The anions that you need to know are:
>
> Group 7 (1– ions)
> chloride, Cl^-
> bromide, Br^-
> iodide, I^-
>
> Compound ions (2– ions)
> carbonate, CO_3^{2-}
> sulfate, SO_4^{2-}

CATION TESTS

1. Clean a flame test wire loop. Dip it into hydrochloric acid and then into a sample of the compound. Hold the wire in the blue part of a Bunsen flame.

- crimson flame → Li^+
- yellow flame → Na^+
- lilac flame → K^+
- red flame → Ca^{2+}
- green flame → Cu^{2+}

no flame colour

2. To a fresh solution add drops of sodium hydroxide solution.

- blue precipitate → Cu^{2+}
- green precipitate → Fe^{2+}
- brown precipitate → Fe^{3+}

white precipitate

3. Add further sodium hydroxide and see if the white precipitate dissolves in excess.

- precipitate dissolves → Al^{3+}
- precipitate does not dissolve → Mg^{2+}

ANION TESTS

4. To a fresh sample of the compound, add dilute hydrochloric acid, and bubble any gas given off through limewater.

- fizzing, limewater goes milky → CO_3^{2-}

no fizzing, limewater remains clear

5. To a fresh solution add dilute nitric acid and silver nitrate solution.

- white precipitate → Cl^-
- cream precipitate → Br^-
- yellow precipitate → I^-

no precipitate

6. To a fresh solution add dilute hydrochloric acid and barium chloride solution.

- white precipitate → SO_4^{2-}

Now try this

A solution of some crystals forms a white precipitate with aqueous sodium hydroxide that does not dissolve when an excess of NaOH is added. A fresh sample of the solution gives a yellow precipitate with dilute nitric acid and silver nitrate solution. The crystals do not produce a coloured flame.
Describe what can be deduced from each test, and give the formula of the crystals.   **(4 marks)**

Extended response – Chemical analysis

You can revise the information for this question, which is about **identifying an ionic compound,** on pages 76 and 77.

Worked example

A sample of a pure ionic compound, **X**, is tested.
X contains **three** ions: two positive ions and one negative ion.
A student performs several tests on solid **X**, or on a solution of **X** in water.
The student's tests and the results are given.

Test	Result
• flame test on solid **X**	• lilac flame
• sodium hydroxide solution added to a solution of **X**	• white precipitate forms
• further sodium hydroxide solution added	• the precipitate dissolves forming a colourless solution
• dilute hydrochloric acid added to solid **X**	• no fizzing
• dilute nitric acid and silver nitrate solution added to a solution of **X**	• no precipitate forms
• dilute hydrochloric acid and barium chloride solution added to a solution of **X**	• white precipitate forms

Use the results to identify the ions contained in **X**. **(6 marks)**

The lilac flame test shows that X contains potassium ions, K$^+$.

Adding sodium hydroxide solution and forming a white precipitate shows that X contains Al^{3+}, Mg^{2+} or Ca^{2+} ions. Only the aluminium hydroxide white precipitate dissolves in excess sodium hydroxide solution, so X must contain Al^{3+} ions.

There is no reaction with either hydrochloric acid or silver nitrate solution, showing that X does not contain carbonate ions or halide ions.

A white precipitate forms with dilute hydrochloric acid and barium chloride solution. This shows that X contains sulfate ions.

X contains K$^+$, Al^{3+} and SO$_4$$^{2-}$ ions.

This answer logically follows the tests in the order in the results table.

You have to know only five flame colours (Li$^+$, Na$^+$, K$^+$, Ca^{2+}, Cu^{2+}), and only one of those is lilac.

The answer needed only to mention Al^{3+}, but giving the extra information about Mg^{2+} or Ca^{2+} could be useful in getting marks if another part of the answer were not complete.

The information about the tests with hydrochloric acid and silver nitrate solution is not needed to identify the ions in **X**, but giving the extra information could be useful in getting marks if another part of the answer were not complete.

Now try this

Plan an experiment to demonstrate that a solid is sodium carbonate. **(6 marks)**

The early atmosphere and today's atmosphere

The Earth's **atmosphere** has stayed much the same for the last 200 million years.

The atmosphere today

The two main gases in the atmosphere are:
• nitrogen (about 4/5)
• oxygen (about 1/5).

There are smaller amounts of other gases and a variable amount of water vapour.

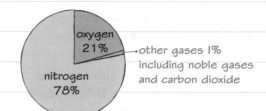
oxygen 21%
nitrogen 78%
other gases 1% including noble gases and carbon dioxide

The early atmosphere

The Earth first had an atmosphere 4.6 billion years ago. This means that:
• there is only limited evidence for scientists
• theories about the early atmosphere have changed and developed.

One theory about the early atmosphere and changes that occurred to the atmosphere is below.

1. The Earth's early atmosphere consisted mainly of carbon dioxide, with little or no oxygen. This is like Mars and Venus today.

first billion years

2. Volcanoes emitted gases.

CO_2 H_2O N_2

Small amounts of CH_4 and NH_3 may also have been produced.

4. Carbon dioxide dissolved in the oceans. Some of this carbon dioxide ended up as precipitates of carbonates, which settled as sediment on the ocean floor.

3. Water vapour from volcanoes condensed to form the oceans.

Worked example

Describe **two** differences between the atmospheres of Earth, and Venus and Mars today. **(2 marks)**

The amounts of nitrogen and oxygen in the Earth's atmosphere are a much higher percentage. The amount of carbon dioxide is much lower.

Gas	Percentage of atmosphere today	
	Venus	**Mars**
nitrogen	3.5	2.7
oxygen	trace	0.1
carbon dioxide	96.5	95.3

Now try this

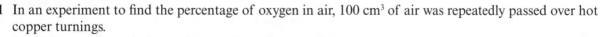

1 In an experiment to find the percentage of oxygen in air, 100 cm³ of air was repeatedly passed over hot copper turnings.
   (a) Balance this symbol equation: ... Cu + O_2 → ... CuO   **(1 mark)**
   (b) The volume at the end was 79 cm³. Calculate the percentage of oxygen.   **(2 marks)**
2 Air can be fractionally distilled. Use data from the table to answer the questions.

Gas	carbon dioxide	oxygen	nitrogen
Boiling point (°C)	−78	−183	−196
Melting point (°C)	−78	−219	−210

   (a) Suggest why carbon dioxide is removed before the air is cooled to −200 °C.   **(2 marks)**
   (b) Suggest why nitrogen leaves the column as a gas but oxygen leaves as a liquid.   **(2 marks)**

Evolution of the atmosphere

About 2.7 billion years ago, algae evolved and **photosynthesis** began. Photosynthesis in algae and in plants formed the oxygen in the atmosphere.

Photosynthesis

As the Earth's atmosphere has developed
• carbon dioxide levels have gone down
• oxygen levels have gone up.

Photosynthesis by plants and algae is one reason for these changes. One billion years after algae first appeared, there was enough oxygen in the atmosphere for animals to evolve.

In photosynthesis:
• carbon dioxide is taken in
• oxygen is given out.

Equation for photosynthesis

$$6CO_2 + 6H_2O \rightarrow C_6H_{12}O_6 + 6O_2$$
carbon dioxide + water → glucose + oxygen

Fossil fuels and limestone

Fossil fuels formed over millions of years from the remains of dead plants and plankton. The carbon they contain originally came from the atmosphere when the organisms were alive. Carbon dioxide was 'locked up' in fossil fuels as:
• carbon in coal
• hydrocarbons in crude oil and natural gas.

Coal is a type of **sedimentary rock**.

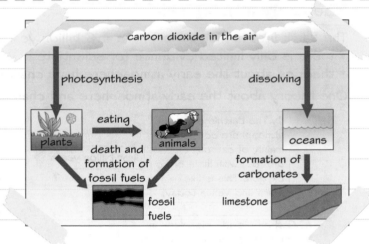

carbon dioxide in the air
photosynthesis   dissolving
plants   eating   animals   oceans
death and formation of fossil fuels   formation of carbonates
fossil fuels   limestone

Worked example

The oceans have an important part to play in absorbing carbon dioxide from the atmosphere. Describe the processes involved.   **(3 marks)**

Carbon dioxide dissolves in the oceans. It forms carbonates, which sea creatures use to make their shells and skeletons. These form limestone rock when the animals die and their shells sink to the seabed. In this way, carbon dioxide becomes locked in sedimentary rocks as calcium carbonate as well as crude oil and natural gas.

Now try this

1 Explain how the Earth's atmosphere has changed because of the evolution of plants and algae.   **(3 marks)**
2 Use the data in the table to help you answer the questions below.

	Temperature (°C)
Boiling point of water	100
Average surface temperature on Venus	460
Average surface temperature on Earth	14

(a) Suggest why Venus has no oceans today.   **(1 mark)**
(b) The Earth's oceans formed around 250 million years after the Earth itself formed. Suggest **one** reason for this.   **(2 marks)**
(c) Suggest **two** reasons why the percentage of carbon dioxide in the atmosphere decreased gradually over 3 billion years.   **(2 marks)**

Greenhouse gases

Greenhouse gases in the atmosphere, including carbon dioxide, methane and water vapour, keep the temperature on Earth warm enough to support life.

The greenhouse effect

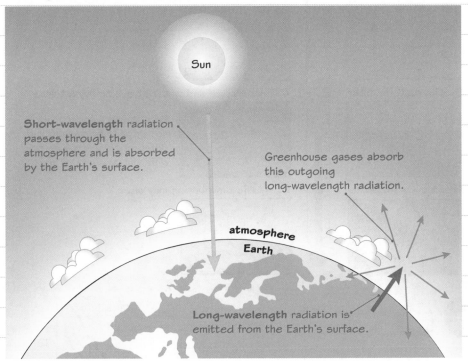

Short-wavelength radiation passes through the atmosphere and is absorbed by the Earth's surface.

Greenhouse gases absorb this outgoing long-wavelength radiation.

atmosphere
Earth

Long-wavelength radiation is emitted from the Earth's surface.

If there were no greenhouse gases in the atmosphere, all of the long-wavelength radiation would escape and the Earth would be too cold to support life.

Human activities increasing the amounts of greenhouse gases

burning **fossil fuels** → CO_2

animal farming → CH_4

deforestation → CO_2

decomposition of rubbish in **landfill** → CH_4

Worked example

Explain how human activities increase the amount of methane in the atmosphere. **(4 marks)**

When rubbish is left in landfill sites bacteria will decompose some of the food waste to release methane gas.

The farming of animals causes an increase in methane. This is because farm animals such as cows release methane when they digest grass. Also, when bacteria decompose animal waste, methane is released.

Now try this

Describe how human activities increase the amount of carbon dioxide in the atmosphere. **(4 marks)**

Global climate change

An increase in the average global temperature is the major cause of climate change.

Potential effects of global climate change

1. Sea level rise

flooding, increased coastal erosion

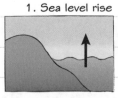

2. Storms more frequent

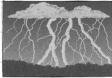

storms may be more severe

3. Rain patterns alter

different parts of the world might get more or less rain, with a pattern of rain each month different to now

4. Temperature and water stress to living things

humans and wildlife may not be able to manage with different temperatures or different amounts of water

5. Food production problems

6. Wildlife distribution may alter

Worked example

Explain how the graph shows evidence for the view that human activity has caused global warming. **(3 marks)**

The human activities of burning fossil fuels and deforestation lead to increasing carbon dioxide in the atmosphere. The graph shows that the rise in average global temperature correlates with the amount of carbon dioxide in the atmosphere.

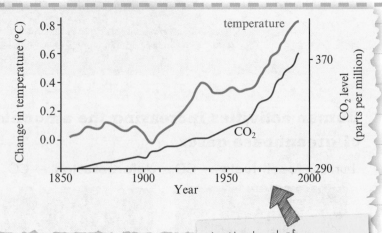

You can revise the effect of human activity on the levels of carbon dioxide and methane, another greenhouse gas, on the previous page. The increase in carbon dioxide levels correlates with the increase in use in fossil fuels.

As the level of carbon dioxide in the atmosphere has risen, so has the average global temperature.

The evidence

Scientists have proposed a **model** to explain how:

- human activities release carbon dioxide
- carbon dioxide is a greenhouse gas
- the increase in average global temperature correlates with the amount of carbon dioxide.

It is difficult to model the Earth's complex climate, but this model, and the evidence for it, has been **peer reviewed**.

The Met Office is where scientists produce information about weather and the climate.

Now try this

Which of the following would **not** be a reason for inaccuracies in a newspaper report about global warming? **(1 mark)**

A   A simplified model has been used.

B   The report contains speculation or opinion.

C   The report may be based on only parts of the evidence.

D   The report used information from the Met Office.

Carbon footprint

The **carbon footprint** is the total amount of carbon dioxide and other greenhouse gases emitted over the full **life cycle** of a product, a service or an event.

Reducing the carbon footprint

Method	Example
use alternative energy sources	Use solar energy instead of burning fossil fuels.
energy conservation	Insulate houses.
carbon capture and storage	Proposal to collect CO_2 given off by burning fossil fuels in power stations and store underground in used oil fields.
carbon taxes and licences	Tax businesses with a carbon footprint to encourage them to reduce their footprint.
carbon off-setting	Paying as part of the cost of an airline ticket for trees to be planted, so the trees absorb as much CO_2 as is given out on the flight.

Carbon neutrality

A carbon neutral process gives off no **net** carbon dioxide in its life cycle. (Any carbon dioxide given out is balanced by carbon dioxide absorbed.)

Worked example

A newspaper article about reducing the carbon footprint is given below.

> Most scientists agree that human activities releasing carbon dioxide and methane cause global climate change, but a few suggest other causes. The climate is very complicated and it is difficult to make predictions of how the climate will change. The public need more information to understand this science, and may not be prepared to change their lifestyles. It may also cost money for individuals and society to make changes that lower the carbon footprint.

(a) Why might some scientists not accept that there is proof that climate change is caused by human activity? **(2 marks)**

There is a **correlation** between carbon dioxide levels and average global temperature, but this does not prove that carbon dioxide **caused** the increase in temperature.

(b) Give **two** lifestyle changes that would reduce the carbon footprint. **(2 marks)**

Eat less meat and cycle to work instead of driving a car.

(c) Explain why all countries have to act together to reduce carbon dioxide amounts. **(2 marks)**

Carbon dioxide goes into the atmosphere and will cause global climate change regardless of the country that emitted the gas.

Life cycle assessments

You can revise more about LCAs on page 91.

The life cycle assessment (LCA) of a product considers all the processes involved in:
- extraction of the raw materials
- transport of raw materials
- manufacture of the product
- transport of the product
- use and operation of the product
- disposal of the waste
- disposal at the end of the product's life.

The carbon footprint of a bunch of flowers from your own garden will be much less than that of a bunch bought in a supermarket that has been packaged and flown in from another country.

Now try this

A consumer buys a plastic bottle of water from the supermarket. Explain why the carbon footprint of this bottle of water may be higher than the carbon footprint of a glass of tap water. **(4 marks)**

Consider the energy required to purify the water, make the containers and transport the water, and any disposal of used materials.

Atmospheric pollution

Coal is mainly carbon. Hydrogen is a fuel. Most other fuels contain both carbon and hydrogen, and many fuels may contain some sulfur. A large range of substances is released when fuels burn, so combustion of fuels is a major source of **atmospheric pollutants**.

Atmospheric pollutants from the combustion of fuels

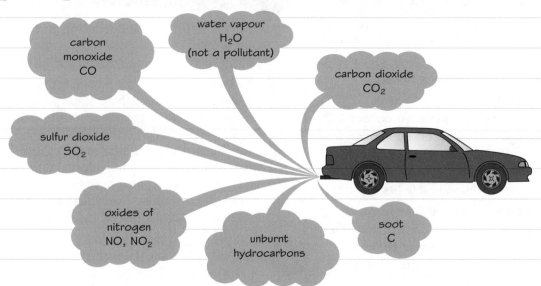

carbon monoxide CO

water vapour H_2O (not a pollutant)

carbon dioxide CO_2

sulfur dioxide SO_2

oxides of nitrogen NO, NO_2

unburnt hydrocarbons

soot C

Worked example

Burning fuels can release carbon monoxide, sulfur dioxide, oxides of nitrogen and particulates.
Describe the problems to human health caused by these substances. **(3 marks)**

- CO is toxic because it combines with haemoglobin in the blood instead of oxygen, preventing enough oxygen being carried round the body.
- SO_2 and oxides of nitrogen cause breathing problems.
- Particulates cause damage to the lungs and so cause breathing problems.

Carbon monoxide is colourless and odourless, so it is difficult to detect.

SO_2 and oxides of nitrogen cause acid rain, which damages plants and buildings.

Particulates in the atmosphere reflect sunlight, reducing the amount that gets to the Earth's surface. This is **global dimming**.

Now try this

1 The table shows the products of combustion of two fuels, A and B.

Fuel	Carbon dioxide	Carbon monoxide	Water vapour	Sulfur dioxide
A	✓	✗	✓	✓
B	✗	✓	✗	✓

(a) Explain which fuel was a hydrocarbon. **(2 marks)**
(b) Explain which fuel underwent incomplete combustion. **(2 marks)**

2 Complete this equation for the complete combustion of propane. **(2 marks)**
$C_3H_8 + ... O_2 \rightarrow ... CO_2 + ... H_2O$

Extended response – The atmosphere

You can revise the information for this question, which is about **pollutant gases in the atmosphere,** on pages 83, 84 and 86.

Worked example

The chart shows car exhaust emissions using biodiesel, a fuel that can be made from plants and diesel.

Explain the problems caused by the emissions of the substances given in the chart, and explain which fuel is preferable using only the information provided. **(6 marks)**

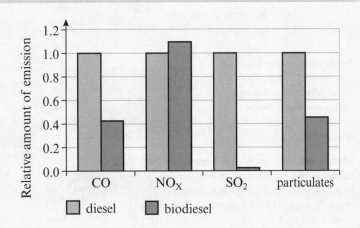

Carbon monoxide combines with haemoglobin in the blood. The haemoglobin is then no longer able to carry oxygen to the cells, so carbon monoxide is toxic.

Oxides of nitrogen cause breathing problems. They dissolve in rainwater to make acid rain. Acid rain causes damage to plants and also erosion of some building materials.

Sulfur dioxide also causes breathing problems and acid rain.

Particulates cause damage to the lungs. They also reflect sunlight, causing global dimming.

Both fuels will release carbon dioxide when they are burned. Biodiesel is produced from plants, so these plants will absorb some carbon dioxide in photosynthesis, and overall release less carbon dioxide.

Biodiesel releases less of every emission except NO_x so is preferable.

The question tells you to use 'only the information provided' so do not try to bring in other factors.

Mention the four emissions one by one to make sure that you cover everything.

 Don't forget to use the information that biodiesel is produced from plants.

 There is only one negative point to biodiesel, which is emission of more oxides of nitrogen.

Now try this

Describe and explain changes in the amount of oxygen from the early atmosphere to today.   **(6 marks)**

Give any relevant equation to help your answer.

The Earth's resources

The Earth's resources are essential for human life. Humans use these resources to provide warmth, shelter, food and transport.

Examples of resources

Plants, such as cotton, and animal skin or hair, such as wool, are farmed to make clothing.

Food grown on Earth by agriculture using fertilisers (see page 97).

Trees occur naturally and are farmed. Timber is used to make buildings. Trees are **renewable** because they can be grown again.

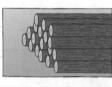

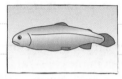

Food occurring naturally in oceans or farmed.

Metals are extracted from **ores** found in the Earth. Iron is extracted and changed into steel and used to make frameworks of buildings (see pages 35 and 94).

Crude oil is a **finite** resource and **industrial processes** such as fractional distillation, cracking and polymerisation are used to manufacture **synthetic** products for example, fuels and plastics (see pages 62, 65 and 70).

Sustainable development

A process or activity is sustainable if:
- it meets our current needs
- it does not compromise the ability of future generations to meet their own needs.

Worked example

The table shows information about the world's copper supplies. Copper is extracted from the Earth and processed to make pure copper and copper alloys, which have many uses.

| known reserves/million tons | 684 |
| demand per year/million tons | 18 |

(a) Explain why copper is described as finite. **(1 mark)**

Once copper resources are used up they cannot be replaced.

(b) Use the table to estimate how many years' supply of copper are known. **(2 marks)**

684/18 = 38 years

(c) Suggest why it is **not** thought that copper supplies will run out. **(2 marks)**

A large amount of used copper is recycled. Also, new copper supplies might be found.

Now try this

1 Read this information about fruit flavours, then answer the question.

> Natural fruit flavours often consist of several different substances. Together they make the distinctive tastes and smell of fruits. It can be difficult and expensive to extract natural flavours to use in food. Esters are used as artificial food flavourings. They are cheap to manufacture but usually a few different substances are mixed to make the flavouring.

Evaluate the use of esters as artificial flavours. Give **one** advantage and **one** disadvantage. **(3 marks)**

2 Ethanol may be used as a fuel on its own or mixed with other fuels such as petrol. It can be made by fermenting food crops such as sugar cane, wheat or maize. Use your scientific knowledge and understanding to evaluate the use of ethanol produced in this way as a fuel. **(4 marks)**

Water

Treating water

Rain is **fresh water** and contains small amounts of dissolved substances.

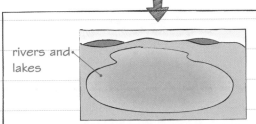
rivers and lakes

Fresh water collects underground and in rivers and lakes.

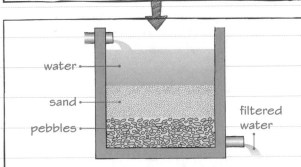
water
sand
pebbles
filtered water

The fresh water is filtered in a **filter bed**.

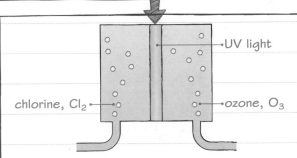
UV light
chlorine, Cl_2
ozone, O_3

The water is **sterilised** to kill microbes using chlorine or ozone or UV light.

Water is essential for life. **Potable** water is water that is safe to drink – it must not have levels of dissolved salts or microbes that are too high.

The UK has plenty of rain, providing all the water that we need. Other countries lack fresh water and use salty water or seawater to make potable water.

seas and lakes

Salty water or seawater can be made into potable water.

Distillation (see pages 3 and 90)	**Reverse osmosis** using a membrane

Desalination is the removal of sodium chloride from salty water or seawater.

Potable water
Households, industry and agriculture produce large amounts of wastewater that must be treated to remove organic matter, microbes and harmful chemicals.

sewage treatment

Screening and removing grit, where large solid particles are removed	sedimentation, where smaller solid particles settle out	anaerobic digestion of **sewage sludge**, where microbes break down solid organic materials in the absence of air	aerobic biological treatment of **effluent**, where microbes break down dissolved organic materials in the presence of air.

Worked example

Explain why potable water is made from salt water only if there are limited supplies of fresh water. **(2 marks)**

Desalination requires large amounts of energy, so this is an expensive way of making potable water. It is cheaper to make potable water from fresh water.

Now try this

Describe the difference between pure water and potable water.
(3 marks)

Practical skills # Core practical – Analysis and purification of water

Water samples can be tested for pH (see page 38) and for the ions present in the water (see pages 76 and 77).

Core practical

Aim

To obtain pure water from a solution of salt in water (salt water).

Apparatus

- eye protection
- distillation apparatus
- Bunsen burner
- salt water

> You should know the names of the distillation apparatus, including the flask and the condenser. See page 3 for information on distillation.

Method

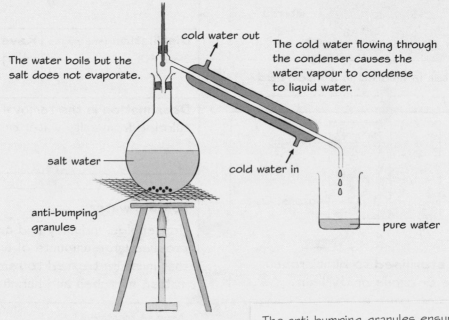

The water boils but the salt does not evaporate.

cold water out

The cold water flowing through the condenser causes the water vapour to condense to liquid water.

salt water

cold water in

anti-bumping granules

pure water

1 Pour the salt water into the round-bottomed flask.
2 Add anti-bumping granules to the flask.
3 Assemble the apparatus and turn on the water to flow through the Liebig condenser.
4 Heat gently.
5 Collect pure water in the beaker.

> The anti-bumping granules ensure smooth boiling and stop the apparatus rattling.

Now try this

Salt water is distilled in the apparatus shown above.
(a) What reading would be shown on the thermometer?   **(1 mark)**
(b) Salt is sodium chloride, NaCl. Describe a test to show that the pure water does not contain chloride ions.   **(3 marks)**

> You can revise the test for chloride ions on page 77.

Life cycle assessment

A life cycle assessment (LCA) can be carried out to assess the total impact on the environment of a product over its whole life, from extracting the raw materials to its disposal.

Factors that can be considered are:

1 use of energy

2 use of raw materials

3 use of water

4 production of waste.

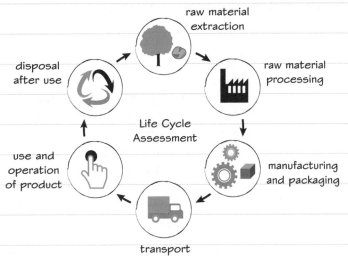

raw material extraction

raw material processing

manufacturing and packaging

transport

use and operation of product

disposal after use

Life Cycle Assessment

Worked example

A government LCA has been conducted on bags used for shopping. The study was carried out, and then reviewed by three independent experts. The findings were compared with three similar studies. A table of results showing only energy use and waste mass, adapted from the study, is shown.

Energy consumption and waste generation per 1000 bags

Bag type	Total energy use (MJ)	Waste mass (g)
single-use plastic bag	22.1	420
plastic bag for life	167.5	170
cotton bag	39.6	1800

 (a) In relative terms, how much more energy does a bag for life use than a single-use plastic bag? **(2 marks)**

$$\frac{167.5}{22.1} = 7.6 \text{ times more energy}$$

(b) In relative terms, how much more waste does a cotton bag produce than a bag for life? **(2 marks)**

$$\frac{1800}{170} = 10.6 \text{ times more waste}$$

 (c) Explain, using **just** this information, why the claim 'cotton bags are better for the environment than plastic bags because they use less energy' is misleading. **(4 marks)**

A cotton bag uses less energy than a bag for life, but more than a single-use plastic bag. However, the total energy use depends on the number of times a bag is reused. A cotton bag used twice uses less energy than two single-use plastic bags. However, the statement ignores the mass of waste produced so the claim 'better for the environment' is misleading because it depends on the factors considered. A cotton bag uses less energy than a bag for life but produces more than 10 times as much waste.

Now try this

 (a) Compare the total energy use and waste mass of the three types of bag used above, for 50 separate single-use plastic bags, 2 plastic bags for life, used 25 times each, and 1 cotton bag, used 50 times. **(4 marks)**

(b) Explain how the government study was checked for reliability. **(2 marks)**

Conserving resources

Many resources are **finite** and their supplies can be conserved by **reduction in use, reuse** or **recycling**. These methods may reduce energy consumption, reduce the amount of waste to be disposed of and reduce our impact on the environment.

Recycling

- Glass bottles can be crushed, melted and made into new glass products.
- Metals can be melted, then made into new metal products.
- Scrap steel can be added to the iron ore in the blast furnace to make more iron.

Reduce

- Packing of food products can be reduced.
- Houses can be insulated and less energy used to heat them.

Reuse

- Glass bottles can be washed and reused.
- A 'bag for life' can be used many times (see page 91).

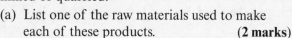

Worked example

Metals, glass, building materials, ceramics and plastics are produced from raw materials that are mined or quarried.

(a) List one of the raw materials used to make each of these products. **(2 marks)**

metals: iron ore or aluminium ore

glass: sand

building materials: limestone

ceramics: clay

plastics: crude oil

> The raw material should be a substance found naturally in the Earth.

(b) Describe some advantages of recycling one of these products. **(3 marks)**

Iron ore has to be mined which is noisy and damages the local environment. Energy is used to manufacture the iron using the iron ore. Recycling scrap iron conserves iron ore supplies, reduces environmental damage at the mine, saves energy and reduces the amount of scrap iron waste that needs to be disposed of.

Now try this

The chart shows seven types of plastic.

PLASTIC RECYCLING CHART						
1	2	3	4	5	6	7
PET	HDPE	PVC	LDPE	PP	PS	OTHER

> Think of the whole process involved from the consumer to the recycling plant. If plastics of different types, 1 to 7 in the chart, are mixed when recycling, then a poor-quality plastic is produced.

Use this information to describe some of the advantages and difficulties involved in recycling plastics. **(4 marks)**

Corrosion

The breaking down of materials by their chemical reaction with substances in the environment, such as oxygen and water, is **corrosion**.

Practical skills — ## Experiment to investigate rusting

The corrosion of iron (or steel) is called **rusting**. This experiment demonstrates what is required for iron nails to rust.

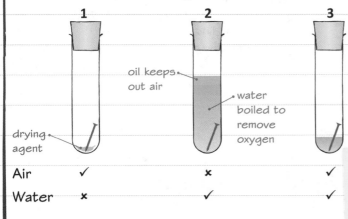

oil keeps out air

water boiled to remove oxygen

drying agent

	1	2	3
Air	✓	✗	✓
Water	✗	✓	✓

Results after being left for one week:

1 no rusting
2 no rusting
3 evidence of rusting

Air **and** water are required for iron to rust. Aluminium naturally has an aluminium oxide coating on its surface, which is impermeable to air or water, preventing corrosion.

Corrosion prevention

There are two ways of preventing corrosion of a metal.

1 Keeping the metal away from water, air or both:
- applying grease to a bicycle chain
- painting an iron/steel bridge
- **electroplating**: covering an object with an unreactive metal using electrolysis, e.g. covering a steel food can with tin.

2 Having a more reactive metal present:
- **sacrificial protection**: magnesium blocks are attached to a steel ship. The magnesium is more reactive than the iron so the magnesium corrodes instead of the steel.
- **galvanising**: covering an object with zinc, e.g. covering an iron bucket with zinc. Initially, the zinc is a barrier to the air and water, but, even if the zinc is scratched, zinc is more reactive than iron so the zinc corrodes instead of the iron.

Worked example

Use the reactivity series to suggest whether sodium, magnesium and copper could be added to an iron nail in the experiment above to provide sacrificial protection.

Increasing reactivity

| sodium |
| calcium |
| magnesium |
| aluminium |
| zinc |
| iron |
| copper |

(5 marks)

Sodium and magnesium are both more reactive than iron. Sodium is so reactive that it would disappear quickly and therefore would not provide sacrificial protection.

Magnesium would corrode instead of iron and slow down iron's rusting.

Copper is less reactive than iron. Iron would act as sacrificial protection for the copper, so iron's rusting would be faster with copper in the test tube.

Steel is an alloy of iron. You can revise alloys on page 94.

Now try this

In the rusting experiment:
(a) explain how the experiment is set up to show that air and water are required for rusting **(4 marks)**
(b) state why, if tube **3** were completely full of water, rusting would still occur. **(1 mark)**

Alloys

Most metals in everyday use are mixtures of metals called **alloys**.

Cast iron

Iron oxide is reduced to iron in a **blast furnace**. Iron straight from the blast furnace is about 96% pure. The impurities it contains make the iron **brittle** and this limits its uses.

Blast furnace iron is used as **cast iron**.

Cast iron is strong in compression and was used to make manhole covers, drain pipes and pillars in buildings. It is now used to make pans and garden furniture.

Steels

Most iron is converted into **steel**. There are different types of steel, but they are all alloys. Steels are mixtures of iron and carbon, often with other metals. The other metals are added in specific amounts to adjust the steel's properties, e.g. chromium and nickel make steel corrosion resistant.

low-carbon steel is softer and easily shaped

high-carbon steel is stronger but brittle

stainless steel is hard and resistant to corrosion

Worked example

Gold alloys are mixtures of gold with copper, silver and zinc, and are used in jewellery. The table gives information about gold–copper alloys.

Gold alloy	Percentage gold	Relative strength	Relative hardness
18 carat	75	4.1	4.4
24 carat	100	1	1

(a) Use information from the table to explain why copper is mixed with gold. **(2 marks)**

to make the alloy stronger and harder than pure gold

(b) Suggest another reason why gold is alloyed with copper. **(1 mark)**

Copper is cheaper than gold.

Pure gold, copper, iron and aluminium are too soft for many uses. For everyday use, they are mixed with small amounts of similar metals to make them harder. For example, copper is mixed with zinc to make brass, used for water taps and door fittings. Copper is mixed with tin to make bronze, used for statues and decorative objects.

Take care to use information given in the question, rather than just repeating it.

Now try this

1 What is an alloy? **(1 mark)**

2 Explain why iron from the blast furnace has limited uses. **(2 marks)**

3 Duralumin is an aluminium alloy. Use the data to evaluate the use of duralumin, aluminium and high-strength alloy steel for aircraft parts. **(4 marks)**

Aluminium and its alloys have a low density.

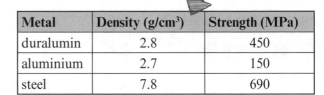

Metal	Density (g/cm³)	Strength (MPa)
duralumin	2.8	450
aluminium	2.7	150
steel	7.8	690

94

Ceramics, polymers, composites

Soda-lime glass
made by heating sand, sodium carbonate and limestone

Borosilicate glass
made by heating sand and boron trioxide

Clay ceramics
made by heating wet clay in a furnace

Borosilicate glass has a higher melting point and withstands changes in temperature better than ordinary (soda-lime) glass.

Composites

Modern composites such as fibreglass use carbon fibres or carbon nanotubes, which are more expensive, but give strong, low-density composites.

Fibreglass has the strength of glass but is not brittle.
- Composites are often **stronger** than their individual components.
- In wood cellulose fibres are held in a lignin matrix.
- Concrete has small stones in a binder of cement and sand.

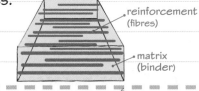
reinforcement (fibres)
matrix (binder)

Thermosoftening polymers

Thermosoftening polymers such as poly(ethene) and poly(propene) soften and melt when heated. Intermolecular forces are weakened when the polymer is warmed, letting the chains slide over each other easily.

tangled but no cross-links
weak intermolecular forces between chains

Thermosetting polymers

Thermosetting polymers have cross-links between them. Thermosetting polymers do not soften or melt when heated.

cross-links

LDPE and HDPE

The properties of polymers depend upon the monomers used and the reaction conditions. Ethene is used to make poly(ethene) and propene makes poly(propene). LDPE – low-density poly(ethene) is made at a different temperature and pressure to HDPE – high density poly(ethene). Different catalysts are also used due to their different properties.

Worked example

The table shows properties of LDPE and HDPE. A plastic becomes too soft to use above its highest useful temperature.

Property	LDPE	HDPE
highest useful temperature (°C)	80	110
relative strength	low	high
flexibility	high	low

Explain which polymer would be best for making plastic drinks cups. **(3 marks)**

HDPE would be best. It is stronger and less flexible than LDPE, which would make a cup easier to hold. Its highest useful temperature is above the boiling point of water, so the cup could hold hot drinks.

Now try this

Experiments have been carried out to test the use of a very thin carbon nanotube composite film as a layer on aircraft wings to de-ice the wings when they get cold. Give the properties of such a composite that make it suitable for this use. **(2 marks)**

The Haber process

Ammonia is used to make ammonium salts for nitrogen-based fertilisers.

It is made in the **Haber process.**

Raw materials

The raw materials for the Haber process are nitrogen and hydrogen:

- **nitrogen** is obtained from the air
- **hydrogen** is obtained from natural gas or other sources. It is also produced by the electrolysis of sodium chloride solution.

The reaction is reversible

nitrogen + hydrogen $\rightleftharpoons$ ammonia

$N_2(g)$ + $3H_2(g)$ $\rightleftharpoons$ $2NH_3(g)$

👎 Only some of the nitrogen and hydrogen react together to produce ammonia.

👎 Ammonia breaks down again into nitrogen and hydrogen.

👍 Unreacted nitrogen and hydrogen are recycled.

The Haber process

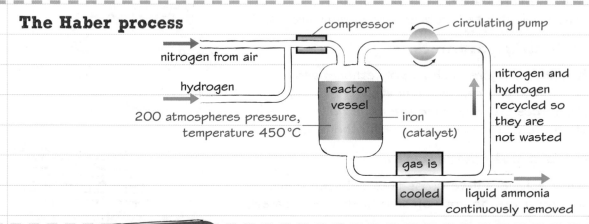

compressor   circulating pump

nitrogen from air

hydrogen

reactor vessel

200 atmospheres pressure, temperature 450 °C

iron (catalyst)

nitrogen and hydrogen recycled so they are not wasted

gas is

cooled   liquid ammonia continuously removed

Worked example

In the Haber process, nitrogen and hydrogen are reacted together to make ammonia, NH_3.

 (a) When nitrogen and hydrogen are left to react in the correct amounts to make ammonia, it is found that after the maximum amount of ammonia has been made, there is still some nitrogen and hydrogen. Explain why. **(2 marks)**

The reaction is reversible, so some ammonia breaks down to reform nitrogen and hydrogen.

(b) What happens to the leftover nitrogen and hydrogen? **(1 mark)**

The gases are recycled back into the reaction so that they are not wasted.

> The forward and backward reactions are going on all the time, so nitrogen and hydrogen are always making ammonia and ammonia is always making nitrogen and hydrogen.

Now try this

Ammonia reacts with sulfuric acid to make ammonium sulfate.

(a) Write a word equation for this reaction. **(1 mark)**

(b) Ammonium sulfate is used as a fertiliser. What is the purpose of a fertiliser? **(1 mark)**

(c) Some fertilisers are described as NPK fertilisers. Apart from nitrogen, name the two other elements found in an NPK fertiliser. **(2 marks)**

> See next page about fertilisers.

> Use your periodic table to help answer this question.

Fertilisers

Crops need compounds of certain elements to grow well, including nitrogen, phosphorus and potassium (NPK). **NPK fertilisers** are used to add these compounds to soil.

Phosphates

Phosphate rocks have various formulae, but all contain the phosphate ion, PO_4^{3-}. They are insoluble, so they are reacted with different acids to make soluble phosphate compounds, which can be used as fertilisers.

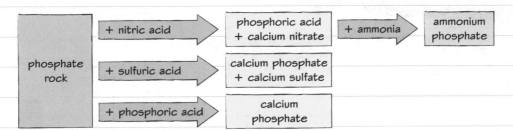

Fertilisers are often spread by spraying a solution of the fertiliser, so the compounds used must be soluble. Potassium chloride (KCl) and potassium sulfate (K_2SO_4) are obtained by mining and are soluble, so they can be used directly as fertilisers.

Worked example

The industrial production of NPK fertilisers can be achieved using several raw materials in an integrated manufacturing process. One part of such a process is the oxidation of ammonia to make nitric acid. Describe how raw materials can be used in an integrated process to make ammonium phosphate. **(6 marks)**

Using the raw materials **nitrogen** and **hydrogen**, ammonia is made in the Haber process. Some of this ammonia is oxidised to make nitric acid.

The raw material **phosphate rock** is reacted with the nitric acid to make phosphoric acid and calcium nitrate.

The phosphoric acid is reacted with more ammonia to make ammonium phosphate.

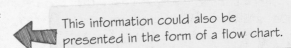
This information could also be presented in the form of a flow chart.

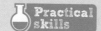
Practical skills In the laboratory, this salt could be made from a solution of ammonia and dilute phosphoric acid.

Now try this

Calculate the percentage by mass of phosphorus in the compound $Ca(H_2PO_4)_2$. **(2 marks)**
(relative atomic masses: H = 1, O = 16, P = 31, Ca = 40)

Maths skills Calculate the relative formula mass, then find the percentage of this that is phosphorus using:
$$\frac{\text{mass of phosphorus}}{\text{relative formula mass}} \times 100\%$$

Extended response – Using resources

You can revise the information for this question, which is about **life cycle assessments**, on page 91.

Worked example

Type of bag	Single-use plastic bag	Single-use bag reused as bin liner
Paper bag	3	7
Bag for life	4	9

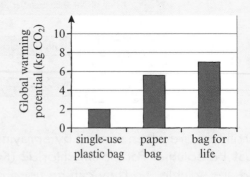

A government LCA has been conducted on bags used for shopping. The table shows the number of times a paper bag or a bag for life have to be used to have the same impact on global warming as a single-use plastic bag.

The graph shows the equivalent amount of carbon dioxide released in the production and disposal of these three types of bag.

Use this information to evaluate which of these types of bag have the least environmental impact. **(6 marks)**

The bag for life gives the largest amount of carbon dioxide released per bag, a little bit more than a paper bag. This is about three times the amount of carbon dioxide released for a single-use plastic bag.

However, these bags may be used more than once. A paper bag causes a higher carbon dioxide release, so it has to be used three times to have the same impact as a single-use plastic bag, or seven times if the single-use plastic bag is then used as a bin liner.

A bag for life causes the highest carbon dioxide release, and has to be used four times to have the same impact as a single-use plastic bag or nine times if the single-use plastic bag is then used as a bin liner.

As bags for life are less likely to be damaged than a paper bag, for example if they get wet, then they should be possible to use more than nine times. If consumers do this, then the bags for life will have the least environmental impact.

> The key to answering this question is to use both sets of data together.

> A single-use plastic bag is designed to be used once, then thrown away. However, if it is used as a bin liner, as many people do, it avoids use of another plastic bin liner, so the overall environmental impact is reduced.

> There are different conclusions to this question, because the overall impact depends on how often consumers reuse bags. As long as the answer is justified, then any of the types of bag could have been chosen.

Now try this

Describe the differences in the ease of making potable water from wastewater, ground water and salt water. **(6 marks)**

> You can revise the information needed to answer this question on page 89.

The Periodic Table of the Elements

Key

relative atomic mass
atomic symbol
name
atomic (proton) number

1	2											3	4	5	6	7	0
					1 **H** hydrogen 1												4 **He** helium 2
7 **Li** lithium 3	9 **Be** beryllium 4											11 **B** boron 5	12 **C** carbon 6	14 **N** nitrogen 7	16 **O** oxygen 8	19 **F** fluorine 9	20 **Ne** neon 10
23 **Na** sodium 11	24 **Mg** magnesium 12											27 **Al** aluminium 13	28 **Si** silicon 14	31 **P** phosphorus 15	32 **S** sulfur 16	35.5 **Cl** chlorine 17	40 **Ar** argon 18
39 **K** potassium 19	40 **Ca** calcium 20	45 **Sc** scandium 21	48 **Ti** titanium 22	51 **V** vanadium 23	52 **Cr** chromium 24	55 **Mn** manganese 25	56 **Fe** iron 26	59 **Co** cobalt 27	59 **Ni** nickel 28	63.5 **Cu** copper 29	65 **Zn** zinc 30	70 **Ga** gallium 31	73 **Ge** germanium 32	75 **As** arsenic 33	79 **Se** selenium 34	80 **Br** bromine 35	84 **Kr** krypton 36
85 **Rb** rubidium 37	88 **Sr** strontium 38	89 **Y** yttrium 39	91 **Zr** zirconium 40	93 **Nb** niobium 41	96 **Mo** molybdenum 42	[98] **Tc** technetium 43	101 **Ru** ruthenium 44	103 **Rh** rhodium 45	106 **Pd** palladium 46	108 **Ag** silver 47	112 **Cd** cadmium 48	115 **In** indium 49	119 **Sn** tin 50	122 **Sb** antimony 51	128 **Te** tellurium 52	127 **I** iodine 53	131 **Xe** xenon 54
133 **Cs** caesium 55	137 **Ba** barium 56	139 **La*** lanthanum 57	178 **Hf** hafnium 72	181 **Ta** tantalum 73	184 **W** tungsten 74	186 **Re** rhenium 75	190 **Os** osmium 76	192 **Ir** iridium 77	195 **Pt** platinum 78	197 **Au** gold 79	201 **Hg** mercury 80	204 **Tl** thallium 81	207 **Pb** lead 82	209 **Bi** bismuth 83	[209] **Po** polonium 84	[210] **At** astatine 85	[222] **Rn** radon 86
[223] **Fr** francium 87	[226] **Ra** radium 88	[227] **Ac*** actinium 89	[261] **Rf** rutherfordium 104	[262] **Db** dubnium 105	[266] **Sg** seaborgium 106	[264] **Bh** bohrium 107	[277] **Hs** hassium 108	[268] **Mt** meitnerium 109	[271] **Ds** darmstadtium 110	[272] **Rg** roentgenium 111	[285] **Cn** copernicium 112	[286] **Uut** ununtrium 113	[289] **Fl** flerovium 114	[289] **Uup** ununpentium 115	[293] **Lv** livermorium 116	[294] **Uus** ununseptium 117	[294] **Uuo** ununoctium 118

* The Lanthanides (atomic numbers 58 – 71) and the Actinides (atomic numbers 90 – 103) have been omitted.

Relative atomic masses for **Cu** and **Cl** have not been rounded to the nearest whole number.

Answers

1. Elements, mixtures and compounds

(a) **C** and **D** (**1**)

(b) **B** and **D** (**1**)

(c) **A** (**1**)

2. Filtration, crystallisation and chromatography

Add water and stir to dissolve the sodium chloride. (**1**)

Filter the mixture. (**1**)

Rinse the solid calcium carbonate in the filter paper and leave to dry. (**1**)

Crystallise the sodium chloride solution that goes through the filter paper. (**1**)

3. Distillation

1 to cool the vapour (**1**); to turn it from the gas state to the liquid state (**1**)

2 first, 78 °C (**1**)

as the ethanol boils off first (**1**)

then rises to 100 °C as the water boils (when all the ethanol has gone) (**1**)

4. Historical models of the atom

In Bohr's model the nucleus was a small, positively charged part of the atom. (**1**)

In the later models of the atom, the nucleus was made of separate protons (**1**) and neutrons (**1**).

5. Particles in an atom

(a) protons = 19 (**1**)

neutrons = 40 − 19 = 21 (**1**)

electrons = 19 (**1**)

(b) They have different atomic numbers (**1**) so different numbers of protons and they are atoms of different elements (**1**).

6. Atomic structure and isotopes

1 Each atom has one electron and one proton (**1**) with no, one or two neutrons (**1**).

The same number of protons means that they are atoms of the same element/hydrogen. (**1**)

2 (a) They have different numbers of neutrons (one with 8 and one with 10 neutrons). (**1**)

(b) They each have 8 protons. (**1**)

They each have 8 electrons. (**1**)

7. Electronic structure

1 (a) 2,4 (**1**)

(b) 2,8,6 (**1**)

(c) 2,8,8,2 (**1**)

2

*13 electrons (**1**), arranged 2,8,3 (**1**)*

8. Development of the periodic table

1 (a) Group 0 (**1**)

(b) Similarity: it contains Li, Na, K, Rb/atoms of the elements can all form 1+ ions. (**1**)

Difference: there are additional elements in Mendeleev's table (H, Cu, Ag). (**1**)

2 Gaps had to be left in the table. (**1**)

Some of the elements in order of atomic weight had to be reversed. (**1**)

9. The modern periodic table

Lv should go into Group 6 (**1**) because its atoms have six electrons in their outer shell (**1**).

10. Group 0

D (**1**)

11. Group 1

1 any temperature above 20 °C and below 98 °C (**1**)

It is lower than sodium (**1**) but above room temperature as potassium is a solid. (**1**)

2 (a) $2Rb(s) + 2H_2O(l) \rightarrow 2RbOH(aq) + H_2(g)$

balancing, (**1**) *state symbols* (**1**)

(b) Same: any two from: effervescence, solid floats, solid moves around, solid disappears (**2**)

Different: Rb would melt/catch fire/react faster (**1**)

12. Group 7

1 solid (**1**)

The melting point increases down the group, so, as iodine is solid, astatine will be too. (**1**)

2 (a) The mixture turns brown. (**1**)

(b) Bromine is more reactive than iodine (**1**) so bromine displaces iodine (**1**).

bromine + potassium iodide → iodine + potassium bromide (**1**)

13. Transition metals

1 high melting point (**1**), good conductor of heat (**1**)

2 (a) low-carbon steel (**1**)

(b) Advantage: lower density so car lighter (**1**) and uses less fuel/accelerates more easily (**1**)

Disadvantage: less strong (**1**) so greater hazard in crash (**1**) OR more expensive (**1**) so car would cost more (**1**)

(c) Titanium gives extra strength (**1**) but would make the car heavier (**1**) and more expensive (**1**) so should be used only when strength is the major consideration, e.g. in an armoured car (**1**).

14. Chemical equations

1 $50 - 22 = 28$ (**1**), $28/2 = 14$ g (**1**)

2 There are two H atoms on each side (**1**) and two F atoms on each side (**1**).

15. Extended response – Atomic structure

Answers may include some of the following points: (**6**)

- A sodium atom has 11 electrons arranged 2,8,1.
- A chlorine atom has 17 electrons arranged 2,8,7.
- An argon atom has 18 electrons arranged 2,8,8.
- When atoms react they often end up with a full outer shell.
- Argon atoms already have a full outer shell so argon is unreactive.
- Sodium atoms have one electron in the outer shell.
- When sodium atoms react they lose the outer electrons, forming Na^+ ions.
- Sodium reacts forming ionic compounds.
- Chlorine atoms have seven electrons in their outer shell.
- When chlorine atoms react, they can gain one electron forming Cl^- ions.
- Chlorine atoms can also react by sharing electrons to gain a full outer shell.
- Chlorine reacts forming ionic compounds and covalent compounds.

16. Forming bonds

(a) magnesium + nitrogen → magnesium nitride (**1**)

(b) Magnesium atoms lose electrons (**1**) which are transferred to nitrogen atoms (**1**) and the oppositely charged ions attract (**1**) forming ionic bonds (**1**).

17. Ionic bonding

(a) Magnesium atoms lose two electrons (**1**) to form Mg^{2+} ions (**1**) and the chlorine atoms gain one electron (**1**) to form Cl^- ions (**1**).

(b) $MgCl_2$ (**1**)

19. Covalent bonding

1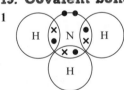
bond showing dot and cross between N and each H (**1**), three hydrogens joined to nitrogen (**1**), lone pair (**1**)

2 (a) The carbon atom and each chlorine atom give an electron (**1**) which is shared (**1**) and there are four covalent bonds (**1**).

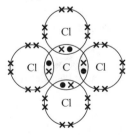

(b) bond showing dot and cross between C and each Cl (**1**), four chlorines joined to carbon (**1**), chlorines each have three lone pairs (**1**)

20. Small molecules

(a) fluorine – gas

chlorine – gas

bromine – liquid

iodine – solid

all correct (**2**), *three correct* (**1**)

(b) Going down Group 7, the halogen molecules increase in size (**1**) so the intermolecular forces increase (**1**) and more energy is required to boil the halogen (**1**).

21. Polymer molecules

A (**1**)

22. Diamond and graphite

(a) The atoms in giant covalent structures are joined with strong covalent bonds (**1**) and a lot of energy is needed to break these bonds to melt the substance (**1**).

(b) C (**1**)

23. Graphene and fullerenes

They are strong (**1**) and have a low density (**1**).

24. Metallic bonding

(a) Copper has highest melting point (**1**) so will not melt (**1**), but other metals would melt/ react with food/ react with water (**1**).

(b) It is a good conductor of electricity. (**1**)

25. Giant metallic structures and alloys

1 The iron and carbon atoms are different sizes (**1**) so the lattice is disrupted (**1**) and the layers in steel cannot slide as easily as in pure iron (**1**).

2 (a) a lattice/giant structure of sodium ions (**1**) surrounded by delocalised electrons (**1**)

The attraction between the ions and the delocalised electrons is metallic bonding. (**1**)

(b) The delocalised electrons can move (**1**) and carry electrical charge and thermal energy/heat (**1**).

(c) The layers can slide over each other (**1**) keeping the metallic bonding intact (**1**).

26. The three states of matter

(a) carbon dioxide: gas; diamond: solid; ethanol: liquid; sodium carbonate: solid (**1**)

(b) Carbon dioxide and ethanol consist of simple molecules (**1**) with only weak intermolecular forces (**1**).

Sodium carbonate has an ionic structure with strong ionic bonds. (**1**)

Diamond has a giant covalent structure with strong covalent bonds. (**1**)

The stronger the forces between the particles the higher the melting point. (**1**)

27. Nanoscience

1 A buckyball molecule is a hollow ball of carbon atoms. A drug molecule could be 'caged' inside one to make a tiny pill. (**1**) Lots of these could be used to get the drug inside the body. (**1**)

2 • The coating is invisible so that you can still see out of the window (**1**) but less heat energy is lost in winter, reducing energy bills (**1**).

28. Extended response – Bonding and structure

Answers may include some of the following points: (**6**)

- Sodium consists of sodium ions in a lattice surrounded by delocalised electrons.
- The metallic bond is between the ions and the sea of electrons.
- Chlorine consists of chlorine molecules – two atoms joined with a strong covalent bond (these bonds are not broken in melting).
- There are weak intermolecular forces between the chlorine molecules.
- Sodium chloride consists of Na^+ ions and Cl^- ions in a lattice.
- The ionic bond is between the oppositely charged ions.

- To melt sodium the metallic bonds must be broken, to melt chlorine the intermolecular forces must be overcome and to melt sodium chloride the ionic bonds must be broken.
- The strengths are ionic bonds > metallic bonds > intermolecular forces.
- So the order of energy required is the same, giving the order of melting points: $NaCl > Na > Cl_2$.

29. Relative formula mass

1 (a) 120 (**1**)
(b) 106 (**1**)
(c) 30 (**1**)
2 (a) 78 (**1**)
(b) 96 (**1**)
(c) 342 (**1**)

30. Balanced equations and masses

(a) **A** (**1**)
(b) **C** (**1**)
(c) **B** (**1**)

31. Concentration of a solution

1 volume = 500/1000 = 0.5 dm^3 (**1**)
concentration = amount in g/volume = 9.8/0.5 (**1**)
= 19.6 g/dm^3 (**1**)
2 **C** (**1**)

32. Reaction yields

1 (a) The actual yield will be lower than the theoretical yield. (**1**)
(b) Other products may have formed. (**1**)
Some product may have been lost when the product was purified. (**1**)
2 percentage yield = 0.9 / 1.2 × 100 (**1**)
= 75% (**1**)

33. Atom economy

(a) 112/214 (**1**) × 100 = 52% (**1**)
(b) 112/244 (**1**) × 100 = 46% (**1**)
(c) Aluminium is extracted by electrolysis (**1**) so it is very expensive (**1**).

34. Reactivity series

The order is **B, A, C**. (**1**)
C is least reactive because it does not fizz with water. (**1**)
B displaces **A** from its chloride so **B** is the most reactive. (**1**)

35. Oxidation, reduction and the extraction of metals

1 (a) Carbon is oxidised (**1**) because it gains oxygen (**1**).
(b) Iron oxide is reduced (**1**) because it loses oxygen (**1**).
2 (a) Potassium reacts with oxygen (**1**) and the oil keeps the potassium away from the air (**1**).
(b) The potassium is silvery because it is a metal (**1**) but when exposed to air it reacts with oxygen (**1**) to form potassium oxide which is dull (**1**).

36. Reactions of acids

(a) $Zn + 2HCl \rightarrow ZnCl_2 + H_2$
reactant formulae (**1**), *product formulae* (**1**), *balancing* (**1**)
(b) The zinc gets smaller/disappears (**1**) and there is fizzing (**1**).

37. Core practical – Salt preparation

Add **excess** copper carbonate to the nitric acid. (**1**)

Filter to remove the excess solid. (**1**)
Heat only until crystals start to form. (**1**)

38. The pH scale

1 (a) red/orange (**1**)
(b) blue/purple (**1**)
(c) green (**1**)
2 The pH meter is more accurate (**1**) and a sample does not need to be taken (**1**).

39. Core practical – Titration

(a) The pipette is more precise. (**1**)
(b) Universal indicator would show a range of colours (**1**) and would not give a sharp colour change (**1**).

40. Extended response – Quantitative chemistry

Answers may include some of the following points: (**6**)
- Method A has atom economy = (2 × 46) / (2 × 46 + 2 × 44) × 100% = 92/180 × 100% = 51%
- Method B has 100% atom economy because there is only one product.
- The waste in method A is carbon dioxide (a greenhouse gas).
- Method B has no waste.
- A method with a high atom economy has less waste.
- A method with a high atom economy may be more economical.

41. Electrolysis

(a) The ions are fixed in a lattice in the solid (**1**) but are free to move in the liquid (**1**).
(b) copper (**1**) chlorine (**1**)

42. Aluminium extraction

(a) melt or dissolve in water (**1**)
(b) sodium + potassium chloride → sodium chloride + potassium
reactants (**1**) *products* (**1**)
$Na + KCl \rightarrow NaCl + K$
reactants (**1**) *products* (**1**)

43. Electrolysis of solutions

	anode	cathode
(a)	bromine (**1**)	hydrogen (**1**)
(b)	chlorine (**1**)	copper (**1**)
(c)	oxygen (**1**)	hydrogen (**1**)

44. Core practical – Electrolysis

(a) Graphite conducts electricity (**1**) and does not react with any of the electrolytes or products (**1**).
(b) Put damp litmus paper into the gas (**1**) and it bleaches/turns white (**1**).
(c) Chlorine is toxic. (**1**)
(d) Some of the chlorine gas dissolves. (**1**)

45. Extended response – Chemical changes

Answers may include some of the following points: (**6**)
- You have equal-sized pieces of each metal.
- You measure out four equal volumes of acid and add to test tubes.
- You add a different metal to each test tube.
- You make observations/monitor the reaction.

- You place the metals in order of most vigorous to least vigorous reaction.
- Potassium and sodium are very reactive so this may be unsafe – use a large container of water instead.
- Copper and gold are very unreactive so you ask a teacher to add the metals to concentrated acid instead.

46. Exothermic reactions

$24.3 - 15.8 = 8.5\,°C$ (**1**)

The reaction is exothermic. (**1**)

47. Endothermic reactions

(a) The solid disappears (**1**) and there is fizzing (**1**).

(b) Measure the temperature of the acid with a thermometer. (**1**)

Add the solid and stir. (**1**)

Take the temperature when the fizzing stops. (**1**)

The temperature falls, showing an endothermic reaction. (**1**)

48. Core practical – Energy changes

(a) Add a lid (**1**) to minimise heat escaping. (**1**)

(b) $25\,°C$ (**1**)

In the original experiment there is an excess of zinc (**1**) so adding more zinc gives no more reaction. (**1**)

49. Activation energy

1   The products energy line is lower than the reactants (**1**) showing that energy would be released to the surroundings (**1**).

2   (a) $2H_2 + O_2 \rightarrow 2H_2O$

formulae (**1**), *balancing* (**1**)

(b) so that some hydrogen and oxygen molecules will have the activation energy (**1**)

3   The line increases showing the energy needed to start a reaction (**1**), which is the activation energy (**1**). It then goes down because the energy of the products is lower than the reactants (**1**) because this is an exothermic reaction (**1**).

50. Cells

The hydrogen is made by the electrolysis of water (**1**) and the electricity needed for electrolysis may be made using fossil fuels. (**1**). The burning of fossil fuel releases carbon dioxide (**1**).

51. Extended response – Energy changes

Answers may include some of the following points: (**6**)

- Measure the sodium hydroxide with a pipette/ measuring cylinder.
- Place it in a polystyrene cup.
- Take the temperature with a thermometer.
- Add the acid $1\,cm^3$ at a time/use a burette.
- Stir and take the temperature for each addition.
- Keep adding the acid until the temperature stops rising.
- Calculate the temperature rise.
- Wear safety goggles.

52. Rate of reaction

(a) $(135.60 - 135.20)/300 = 0.0013\,g/s$ (**1**)

(b) As the reactants are being used up, the mean rate of reaction falls. (**1**)

53. Rate of reaction on a graph

average rate $= 132/2$ (**1**)

$= 66\,cm^3/min$ (**1**)

54. Collision theory

If the concentration of the acid is increased there will be more reacting particles (**1**) so the frequency of collisions between acid particles and magnesium will be increased (**1**) and the rate of reaction will increase (**1**).

55. Rate: pressure, surface area

pressure higher/more molecules (**1**)

collision frequency higher (**1**)

rate of reaction higher/time to form HI reduced (**1**)

three times as many molecules, so would expect rate to be three times higher (**1**)

56. Rate: temperature

1   (a) The rate of reaction increases (**1**), the surface area to volume ratio of the marble increases (**1**) and so the frequency of collisions increases (**1**).

(b) The rate of reaction decreases, (**1**) the acid particles become less crowded/there are fewer reactant particles in the same volume (**1**) and so the frequency of collisions decreases (**1**).

(c) The rate of reaction increases (**1**), the (acid) particles move around faster/have more energy (**1**), more collisions have the necessary activation energy or higher energy (**1**) and so the frequency of successful collisions increases (**1**).

2   Increase the concentration of the acid. (**1**)

Increase the temperature of the acid. (**1**)

Use magnesium powder (i.e. increase the surface area). (**1**)

57. Core practical – Rate of reaction

(a) As the concentration increases, the time falls. (**1**)

(b) $120–160$ (**1**) s (**1**)

58. Catalysts

Measure in a gas syringe the oxygen gas given off as a measured volume of hydrogen peroxide solution decomposes. (**1**)

Record the time for a fixed amount of gas to be produced. (**1**)

Repeat the same experiment three times, first with some **A** added, then the same amount of **B**, and finally the same amount of **C**. (**1**)

If the reaction with any of the added solids produces the same volume of gas in a much shorter time, then the added solid is a catalyst. (**1**)

59. Reversible reactions

The hydrated copper sulfate loses water (**1**) and becomes anhydrous copper sulfate, which is white (**1**).

If anhydrous copper sulfate is added to a liquid, it will turn blue if water is present (**1**) because the dehydration reaction is reversed (**1**).

60. Equilibrium

C (**1**)

61. Extended response – Rates of reaction

Answers may include some of the following points: (**6**)

- A is small chips.
- B is medium chips.
- C is large chips.
- The small chips have a larger surface area.
- The chips with the larger surface area react faster.
- This is because there are more frequent collisions.
- The faster reaction has a steeper graph/ gives off all of the gas faster/ gives off all the gas in a shorter time.
- In each experiment, the mass of the chips was the same so the same amount of gas was given off.

62. Crude oil

The higher the number of carbon atoms in a molecule, the higher the boiling point. (1)

63. Alkanes

1 (a) a molecule with single bonds only (1)

   (b) a molecule containing carbon and hydrogen only (1)

2 C_nH_{2n+2} (1)

3 The formula matches the general formula of alkanes: with five carbon atoms there are $(2 \times 5) + 2 = 12$ hydrogen atoms. (1)

4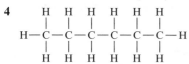

molecule with 6 carbons and 14 hydrogens (1), *drawn correctly* (1)

64. Properties of hydrocarbons

1 The liquefied petroleum gas molecules have fewer carbon atoms (1) so they have a lower boiling point (1) and condense at a cooler temperature/higher position in the tower (1).

2 (different) boiling point (1)

3 Hot crude oil passes into a fractional distillation tower (1); the different fractions condense at different places in the tower (1).

65. Cracking

(a) to evaporate the paraffin (1)

(b) thermal decomposition (1)

(c) Y (1)

66. Alkenes

(a) B, C, D (1)

(b) B, D (1)

(c) C (1)

67. Reactions of alkenes

(a)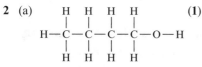

   (3) *each formula* (1)

(b) diiodopropane (1)

(c) Mix each sample with bromine water. (1)
   The mixture with propane stays orange. (1)
   The mixture with propene turns colourless. (1)

68. Alcohols

1 Add a few drops of universal indicator solution. (1)
   Indicator goes red only in the ethanoic acid. (1) *(Allow other named indicators with relevant colours.)*

2 (a)

   H—C—C—C—C—O—H (1)

   (b) solvent/fuel (1)

3 $6CO_2 + 8H_2O$
   water as other product (1), *balancing* (1)

69. Carboxylic acids

1 D (1)

2 (a) The functional group (1) is COOH (1).

   (b) ester (1)

70. Polymers

1 (a)

   (1)

   (b) poly(chloroethene) (1)

2 It is saturated (1) because the double bonds are removed during polymerisation (1).

71. DNA

C (1)

72. Extended response – Organic chemistry

Answers may include some of the following points: (6)

- Add sodium carbonate to a sample of each liquid.
- There will be fizzing in the sample that is a carboxylic acid.

OR

- Add litmus paper/named indicator to a sample of each liquid.
- The litmus paper will turn red/indicator changes to acidic colour in the sample that is a carboxylic acid.
- Add bromine water to a sample of each liquid.
- The mixture will turn colourless in the sample that is an alkene.
- Add a small piece of sodium to a sample of each liquid.
- There will be fizzing in the samples that are the alcohol and the carboxylic acid.

73. Pure substances and formulations

(a) The paint is a mixture (1) in which each component has a different use (1) and the components are added in measured quantities (1).

(b) *Any two of:* (2)

   pigment: gives the colour of the paint

   brightener: makes the colour brighter

   glue: binds the parts of the paint together

   thickener: to stop the paint running off/give better coverage

   additives: to allow components to mix/help paint to retain moisture/stop fungal growth

   water: dissolves all the components

74. Core practical – Chromatography

Any three errors and how to improve that error from the following: (6)

Error	Improvement
an ink pen used for start line	use a pencil
start line under the solvent	chromatography paper dips into the solvent, with the start line above the solvent
solvent rises to top of paper	take paper out before solvent reaches top
$R_f = \dfrac{\text{distance moved by substance}}{\text{length of paper}}$	$R_f = \dfrac{\text{distance moved by substance}}{\text{distance moved by solvent}}$

75. Tests for gases

The gas is hydrogen (1) which burns with a squeaky pop, making water (1) that is seen as condensation (1).

76. Tests for cations

(a) In a flame test (1) calcium ions give a red flame (1). Adding sodium hydroxide solution to a solution containing the ions (1) forms a white precipitate (1).

(b) When adding sodium hydroxide solution (**1**) both ions give a white precipitate. (**1**)

77. Tests for anions

1 The rainwater does not contain chloride ions. (**1**)

The seawater and the damp contain chloride ions. (**1**)

The dampness was caused by seawater. (**1**)

2 Add some sodium hydroxide to samples of the solutions from each bottle. (**1**)

The one that forms a white precipitate and dissolves in excess is aluminium sulfate. (**1**)

The one that forms a white precipitate and does not dissolve in excess is magnesium sulfate. (**1**)

To a solution of each of the other two add dilute nitric acid and silver nitrate solution. (**1**)

The one that forms a cream precipitate is potassium bromide. (**1**)

The one that forms a yellow precipitate is potassium iodide. (**1**)

78. Flame emission spectroscopy

(a) Solutions of both compounds give a white precipitate with sodium hydroxide solution (**1**) but the calcium ions and the magnesium ions have unique/different line spectra (**1**).

(b) Lithium ions and calcium ions both give flames that are a shade of red (**1**) so they could be confused (**1**) and, if impure, the colours could be masked (**1**), but flame emission spectroscopy gives unique/different line spectra (**1**).

79. Core practical – Identifying a compound

NaOH test: the crystals contain Ca^{2+} or Mg^{2+} ions (**1**)

$AgNO_3$ test: the crystals contain I^- ions (**1**)

flame test: the crystals do **not** contain Ca^{2+} ions so must contain Mg^{2+} ions (**1**)

formula: MgI_2 (**1**)

80. Extended response – Chemical analysis

Answers may include some of the following points: (**6**)

- Add dilute hydrochloric acid.
- Pass the gas given off through limewater.
- The limewater turns milky.
- This proves that carbonate ions are present.
- Clean a flame test wire loop in hydrochloric acid.
- Dip wire loop in acid.
- Dip in solid.
- Hold in blue part of flame.
- Yellow flame produced.
- This proves that sodium ions are present.
- Wear safety goggles.

81. The early atmosphere and today's atmosphere

1 (a) $2Cu + O_2 \rightarrow 2CuO$ (**1**)

(b) volume oxygen = $100 - 79 = 21 \, cm^3$ (**1**)

percentage = $21/100 \times 100 = 21\%$ (**1**)

2 (a) Carbon dioxide's melting point is above $-200\,°C$ (**1**) so it would form a solid (**1**).

(b) Temperature must be above $-196\,°C$ (**1**) but below $-183\,°C$ (**1**).

82. Evolution of the atmosphere

1 Photosynthesis by plants and algae (**1**) reduces the percentage of carbon dioxide (**1**) and increases the percentage of oxygen. (**1**)

2 (a) It is too hot for water to condense. (**1**)

(b) The Earth's surface had to cool below $100\,°C$ (**1**) and it took this long for the cooling to happen (**1**).

(c) It took time for the oceans to form and then absorb carbon dioxide. (**1**)

Photosynthesis began only after algae and plants evolved. (**1**)

83. Greenhouse gases

Fossil fuels, when burnt, release carbon dioxide (**1**), e.g. burning coal in power stations/burning petrol in cars (**1**). Deforestation (**1**) means that the cut-down trees cannot absorb carbon dioxide in photosynthesis (**1**).

84. Global climate change

D (**1**)

85. Carbon footprint

Tap water requires energy for purification, but little energy to transport it through pipes. (**1**)

Bottled water may require less energy for purification (**1**) but requires fuel to deliver the bottles to shops (**1**). The plastic bottles require energy to make and may also add to the carbon footprint if they are not recycled (**1**).

86. Atmospheric pollution

1 (a) fuel **A** (**1**) because only **A** produces water (which comes from the oxidation of the hydrogen in the hydrocarbon) (**1**)

(b) fuel **B** (**1**) because carbon monoxide is produced (**1**)

2 $C_3H_8 + 5O_2 \rightarrow 3CO_2 + 4H_2O$

carbon dioxide and water numbers (**1**), *oxygen number* (**1**)

87. Extended response – The atmosphere

Answers may include some of the following points: (**6**)

- The early atmosphere contained little or no atmosphere/oxygen.
- Algae and other plants evolved.
- Photosynthesis began.
- Photosynthesis involves taking in carbon dioxide and releasing oxygen.
- carbon dioxide + water $\rightarrow$ glucose + oxygen
- $6CO_2 + 6H_2O \rightarrow C_6H_{12}O_6 + 6O_2$
- Oxygen levels built up to the amount that they are today.
- The percentage is 21%.

88. The Earth's resources

1 *Any one advantage, for example:*

Esters are cheap to make/it is easier to make esters than to extract natural flavours/a lot of artificial flavouring may be made. (**1**)

Any one disadvantage, for example:

Only a few different substances are used so the flavour might not be natural/some people do not like eating artificial substances. (**1**)

Justified opinion, for example:

I think artificial food flavours are fine to use because they make the food cheaper/I don't think artificial food flavours should be used because it means that the food is not natural. (**1**)

2 *Two advantages for 1 mark each, for example:*

Ethanol is a renewable resource. (**1**)

Overall it produces less carbon dioxide when burned than fuels such as petrol. (**1**)

Two disadvantages for 1 mark each, for example:

It takes a long time to grow crops. (**1**)

The crop plants used to make the ethanol could have been used to feed people instead. (**1**)

89. Water

Pure water contains water only with no other substances. (1)

Potable water contains dissolved substances (1) but at a low enough level to make the water safe to drink. (1)

90. Core practical – Analysis and purification of water

(a) 100 °C (1)

(b) Add dilute nitric acid (1) then silver nitrate solution (1).

No (white) precipitate forms/liquid remains clear. (1)

91. Life cycle assessment

(a)

Bag type	Total energy use (MJ)	Waste mass (g)
single-use plastic bag	22.1 × 50 = **1105**	420 × 50 = **21 000**
plastic bag for life	167.5 × 2 = **335**	170 × 2 = **340**
cotton bag	39.6 × 1 = **39.6**	1800 × 1 = **1800**

energy use values (1), *waste mass values* (1)

The single-use plastic bag uses more energy and creates more waste. (1)

The bag for life creates less waste but uses more energy than the cotton bag. (1)

(b) Other scientists checked the work (1) and the findings were compared with other studies (1).

92. Conserving resources

Any two advantages from: (2)

Plastics are made from crude oil.

Crude oil reserves are conserved.

The waste plastic does not need to go into landfill.

Plastics use up space in landfill because most of them do not easily break down.

Any two difficulties from: (2)

The consumer has to make an effort to recycle the plastic.

The different plastic types then have to be sorted.

Energy is used transporting the waste plastics, and also cleaning and sorting them.

93. Corrosion

(a) Tube 1 has a drying agent to remove any water, so it has air only. (1)

Tube 2 has boiled water, which removes any air, and oil to stop any more air getting in. (1)

Tube 3 has air and water present. (1)

Only the nail in tube 3 rusts, showing that air and water are required for rusting. (1)

(b) Air is dissolved in the water. (1)

94. Alloys

1 a mixture of metals/of a metal and another element (1)

2 Iron from the blast furnace contains (carbon) impurities (1) which make it brittle (1).

3 Steel is strong but its density is high so the parts would be too heavy. (1)

Aluminium has a lower density than duralumin. (1)

Duralumin has a slightly higher density but three times the strength of aluminium. (1)

Duralumin is the best choice because the aircraft parts will have a low enough mass but high strength. (1)

95. Ceramics, polymers, composites

The composite has low density/would not make the aircraft wing much heavier (1) and the carbon nanotubes can conduct electricity (1).

96. The Haber process

(a) ammonia + sulfuric acid → ammonium sulfate (1)

(b) to make plants grow more (1)

(c) phosphorus (1) potassium (1)

97. Fertilisers

$M_r = 40 + (1 × 4) + (31 × 2) + (16 × 8) = 234$ (1)

percentage P = 62/234 × 100 = 26% (1)

98. Extended response – Using resources

Answers may include some of the following points: (6)

Ground water

- Some countries have plentiful supplies of ground water.
- This is found underground and in lakes and rivers.
- The water has to be filtered and sterilised.
- This is a reasonably straightforward and inexpensive process.

Salt water

- The seas have almost unlimited water supplies.
- The water is desalinated, removing salt.
- This is done by distillation or reverse osmosis.
- These processes use large amounts of energy.
- This makes the processes expensive.
- This method is used only where there is a shortage of ground water.

Wastewater

- Wastewater must have microorganisms and harmful chemicals removed.
- It is screened, sedimentation occurs and then there is treatment by microbes.
- This method is used only where there is a shortage of ground water.

Your own notes

Your own notes

Your own notes

Published by Pearson Education Limited, 80 Strand, London, WC2R 0RL.

www.pearsonschoolsandfecolleges.co.uk

Text © Pearson Education Limited 2017
Typeset and produced by Phoenix Photosetting
Illustrated by Barking Dog Art
Cover illustration by Miriam Sturdee

The right of Mark Grinsell to be identified as author of this work has been asserted by him in accordance with the Copyright, Designs and Patents Act 1988.

First published 2017

20 19 18 17
10 9 8 7 6 5 4 3 2 1

British Library Cataloguing in Publication Data
A catalogue record for this book is available from the British Library

ISBN 978 1 292 13127 6

Printed in Slovakia by Neografia.

Acknowledgements
Content written by Nigel Saunders is included in this book.

The publisher would like to thank the following for their kind permission to reproduce their photographs:

(Key: b-bottom; c-centre; l-left; r-right; t-top)

Alamy Stock Photo: sciencephotos 76r; **Getty Images**: Wf Sihardian / EyeEm 84bl; **Science Photo Library Ltd**: 76l; **Shutterstock.com**: abutyrin 84br, Vladimir Melnik 84cr

All other images © Pearson Education